MEGALODONS, MERMAIDS, AND CLIMATE CHANGE

MEGALODONS, MERMAIDS, AND CLIMATE CHANGE

Answers to Your Ocean and Atmosphere Questions

ELLEN PRAGER AND DAVE JONES

Columbia University Press
New York

Columbia University Press
Publishers Since 1893
New York   Chichester, West Sussex
cup.columbia.edu

Copyright © 2024 Ellen Prager and Dave Jones
Illustrations copyright © 2024 Alece Birnbach
All rights reserved

Library of Congress Cataloging-in-Publication Data
Names: Prager, Ellen J., author. | Jones, David (Meteorologist), author.
Title: Megalodons, mermaids, and climate change : answers to your ocean and atmosphere questions / Ellen Prager and Dave Jones.
Description: New York : Columbia University Press, [2024] | Includes bibliographical references and index.
Identifiers: LCCN 2024014788 | ISBN 9780231212489 (hardback) | ISBN 9780231212496 (trade paperback) | ISBN 9780231559416 (ebook)
Subjects: LCSH: Ocean—Miscellanea. | Atmosphere—Miscellanea. | Ocean—Popular works. | Atmosphere—Popular works. | Climatic changes—Popular works. | LCGFT: Trivia and miscellanea.
Classification: LCC GC21 .P69 2024 | DDC 551.46—dc23/eng/20240529
LC record available at https://lccn.loc.gov/2024014788

Printed in the United States of America

Cover design: Milenda Nan Ok Lee
Cover photo: NASA Earth Observatory images. Taken by Joshua Stevens and Lauren Dauphin, using Landsat data from the U.S. Geological Survey and MODIS Data from LANCE/EOSDIS Rapid Response
Cover art: Shutterstock

Questions are the lifeblood of learning.

CONTENTS

Note from the Authors   ix

1   The Deep Vast Sea   1
2   Dangerous Marine Life   15
3   Jellyfish   31
4   Other Sea Creatures   38
5   Coral Reefs   49
6   Supernatural, Suspicious, or Science   62
7   Lightning   81
8   Hurricanes   90
9   Weather Forecasting and Extreme Events   110

CONTENTS

10 Climate Change   135

11 The Sun   160

12 Information Mixology   176

13 Show Us the Data   183

14 Questions Anyone? Anyone?   188

Acknowledgments   191
Sources and Additional Information   195
Index   217

NOTE FROM THE AUTHORS

WE, THE AUTHORS, Ellen and Dave, are proud science geeks. We are endlessly curious about the planet and passionate about observing, learning, and questioning—science. We sometimes disagree about which is more important, the ocean (Ellen, the marine scientist) or atmosphere (Dave, the meteorologist). The atmosphere holds the oxygen we need to breathe (the meteorologist). But phytoplankton in the ocean produce about half the oxygen in the atmosphere (the marine scientist). Don't get us started, though the schtick is good. We love to laugh (often at ourselves) and use humor to engage audiences and make learning fun. Cumulatively, we've done hundreds of public presentations (on stage, on air, and virtually) and enjoy interacting with scientists, nonscientists, kids, educators, the media, our colleagues, our families, our neighbors . . . Okay, just about anyone.

NOTE FROM THE AUTHORS

Sometimes scientists don't agree on what's more important.

Throughout the years, we've been asked many questions by all sorts of people and audiences. Some questions are asked frequently; others are more unusual, infrequent, or plain old wacky. But even the most cornball, outside-the-box question opens the door to conversation and learning. And so, to promote improved understanding and combat misinformation, we decided to write this book and provide the answers to some of the most frequently asked and zaniest questions we and our colleagues get about the ocean, marine life, weather, climate change, and more. We hope it will also provide a few laughs (especially Alece's fabulous illustrations) and help to

NOTE FROM THE AUTHORS

build a better appreciation for science and the world around us. We also hope that, after reading the book, you will have a better grasp on where to find science-based information about Earth, the atmosphere, and the ocean. Before we dive, or fly, into the book, here's a little about us and a few pivotal moments in our lives that inspired our passions for the sea, the weather, and science.

ELLEN PRAGER

During talks, I am frequently asked what inspired me to become a marine scientist. There's a lot to choose from, including as a kid watching nature-based television shows like Jacques Cousteau and *Wild Kingdom* (I later met and became friends with the wonderful Jim Fowler) and an early snorkeling vacation with my family. But several moments and experiences stand out as especially inspirational on my journey to becoming a marine scientist.

During summers in high school, I worked as a lifeguard and water safety instructor. One day, another lifeguard brought scuba gear to the pool and asked if I'd like to try it. I quickly jumped into the pool with a tank on and reveled at being able to stay underwater beyond my breath-holding capability. I later rushed home and announced that I wanted to get scuba certified. My parents always encouraged me to pursue my dreams through education and hard work. So, using my own funds, I paid for and took a scuba certification course, doing the required open-water dives in the cold, murky, surging waters off the Massachusetts coast. I loved it. In hindsight, this moment led to another life-changing experience: as an undergraduate, I took a semester away from Wesleyan University to study tropical marine science at the West Indies Laboratory in St. Croix, U.S. Virgin Islands.

For years, the West Indies Laboratory was a haven for students and researchers studying coral reefs. Classroom learning was combined with field experience through diving, snorkeling, and independent projects. My professors had jobs I dreamed of, and they quickly became role models and mentors (and later valued friends and colleagues). During the semester, we took a field trip to Hydrolab, a small undersea habitat where scientists lived underwater for a week to study coral reefs. I was so captivated by the program that the following Saturday, I rode my bike across the not-so-flat island to the Hydrolab base to inquire about summer diving jobs.

After confirming that I was an experienced, certified diver who could carry (barely) a twin steel scuba tank rig in each hand, I was hired. Returning to the West Indies Lab, I was exhausted and elated. The other students questioned how in the world I got the job. My reply: I asked. It was not only the starting point of an amazing summer of diving and learning but also an important lesson in being proactive and asking for opportunities—an attitude that has borne fruit throughout my life and career.

Working as a support diver for Hydrolab brought another influential moment in my career. Support or safety diver was essentially a euphemism for underwater "gofer." During missions we brought up empty tanks, filled them, and dove them back down to the undersea lab. We ferried supplies and meals down, assisted in scientific research, tracked divers outside the lab, and were the cleanup crew after missions. It was the best job I could have imagined. One mission, in particular, helped to overcome doubts that I had what it takes to be a "scientist."

The goal of the mission was to quantify the role of parrotfish grazing in nutrient recycling in coral reefs. Coral reefs tend to grow in nutrient-poor water, and parrotfish were known to play an important role, but how big a role was unclear. So, each night a team of scientists left the undersea laboratory, swam to the nearby tank racks, and put on their gear. Using special collecting equipment (plastic baggies), they followed parrotfish about the reef to capture their excrement for later analysis onshore. That's when it hit me: If collecting parrotfish poop was science, I could do it.

The West Indies Lab and Hydrolab were just the beginning of a career filled with amazing opportunities, adventure, humor, and learning—from teaching oceanography aboard tall sailing ships for Sea Education Association, operating out of Woods Hole, Massachusetts, to running a small marine laboratory on a remote island in the Bahamas, doing research in Florida Bay for the U.S. Geological Survey, working in the Galapagos Islands, and coming full circle from a support diver at Hydrolab to living underwater as an aquanaut and later being the chief scientist at its successor, the Aquarius Reef Base. My overall perspective was broadened by serving as the assistant dean at the University of Miami's Rosenstiel School of Marine and Atmospheric Science, while working with the U.S. Commission on Ocean Policy, and as an adviser for federal ocean agencies on the Ocean Research and Resources Advisory Panel.

NOTE FROM THE AUTHORS

Somewhere along the way, I discovered a new passion: communicating science to nonscientists. This led to the publication of children's illustrated books, two series of humor-laced eco-adventure novels for middle graders (so much fun to write!), magazine articles, and popular science books, along with many speaking appearances. Somehow I also became a sought-after expert for television (a lot of learning from your mistakes was involved) and have appeared on CNN, the Weather Channel, NBC, the *Today* show, *Good Morning America*, *CBS Mornings*, the Discovery Channel, and more. And, in what most young audiences consider my best most impressive job so far, I was a consultant for Disney's 2016 movie *Moana*. It's been a great ride that continues today, and I can't wait to see what comes next.

DAVE JONES

While Ellen was immersed in the ocean, I had my head in the clouds. At six years old, I was already wondering why a big dark cloud was in one spot and not in another. Why did rain in the forecast sometimes pour at the end of the street and sometimes not at all? But maybe the most influential and electrifying event occurred when a lightning bolt struck the tree across the street from my parents' house (and my room). After the flash, a glowing ball of bluish-white light floated across the street toward my window before disappearing into thin air (more about lightning in chapters 6 and 7). I'll never forget it! I was mesmerized and wanted to learn everything I could about lightning and weather.

NOTE FROM THE AUTHORS

At age eleven, I was so enthralled with the weather that each day after school I would walk (sometimes run) down the street to the local community college in Catonsville, Maryland. I'd end up in the science building at the weather map wall. They had an awesome Introduction to Meteorology course with a small weather office and a teletype machine that printed out weather bulletins and observations at a back-then crazy speed of 300 baud! That's about thirty characters per second, and the wet paper weather maps had to dry before being hung up.

The head of the Earth Science Department took note of my enthusiasm for those weather maps and my questions to anyone within earshot. One day I packed some old maps into my bag (with permission, of course) so I could rehang them in my home "weather office," an old cleaned-out darkroom. Looking on, the department chair asked if I wanted to take the course. He explained that I could be admitted as a "special student," and since my mother worked at the college, if I passed the class, she could get reimbursed. Like Ellen, my enthusiasm and willingness to ask questions led to my first science (atmospheric) opportunity—a college course at twelve years old.

A dental appointment and my early college education led to another opportunity—my first consulting gig. When my dentist learned I was taking a college meteorology course while in seventh grade, he contacted his buddies who were taking flying lessons. All three of them were having difficulty with the weather portion of pilot training. I ended up teaching them aviation meteorology in the private dining room at a local Chinese restaurant. I was in heaven—talking about weather and aviation while eating as much egg drop soup and chicken with broccoli as possible!

NOTE FROM THE AUTHORS

In community college, I signed up for a noncredit course offered in the evening called Weather Forecasting. My enthusiasm and requests for weather maps (again) after each class got the attention of the professor, Dr. Robert Atlas, who also worked at NASA's Goddard Space Flight Center (GSFC). He asked where I was going for my four-year degree. I had already applied to Penn State because they had one of the best meteorology programs in the nation. He explained that he hires interns to work at NASA, but they have to go to the University of Maryland in College Park. No hard decision there. Without blinking, I replied, "I can go to Maryland!"

My college career was also full of incredible opportunities. While working at GSFC supporting Dr. Atlas's research, I visited the NBC television station in Washington, DC, with a classmate who was an intern. I was thrilled when I got to meet the chief meteorologist, Bob Ryan—the most-watched broadcast meteorologist in the DC area. After a few weeks of enthusiastically observing the fast pace of local news, I was offered a job helping to support their weather office. That made two awesome weather jobs, in addition to my true moneymaking weekend gig at Hecht's (a department store), where, at the mall entrance, I sold credit card applications.

After graduation, I worked for weather graphics companies as well as helped a start-up in the Pacific Northwest that built an automated system to provide windsurfers information on where to find the best wind in the Columbia River gorge.

A call from Bob Ryan brought another unexpected and exciting opportunity. He asked if I wanted to put together a sample tape of me doing a weathercast. Whaat? Who wouldn't want to?

NOTE FROM THE AUTHORS

Just two weeks after my "audition," I was hired by NBC4 in Washington, DC, to launch their new weekend morning show. Wow! I spent nearly ten years at NBC4 and had some amazing experiences, including leading a great team to win a NASA grant to use this thing called the "internet" to access NASA's Earth and space data and put it on TV. In 1995 we built WeatherNet4, the first TV station weather website in the nation. We were the number one TV station in the market, bringing earth and space science data to millions of viewers and had special web pages for educators and students. Several times I was flown to New York City to do the weather for the *Today* show but had to hightail it back to DC to deliver the evening forecast. It was hectic, but crazy fun.

I left NBC4 and ventured out on my own to start a company, StormCenter Communications Inc. I wanted to further pursue innovation in how we use and access data and am especially proud of the patented new technology we developed called GeoCollaborate®. It enables real-time synchronous collaboration across any device or platform to improve the sharing of disparate trusted data for people who need it, such as emergency responders and managers; local, state, or regional authorities; companies; government agencies; planners; decision-makers; and the public. I also work closely with the broadcast meteorology community to provide access to unique data and science content to deliver to the public.

■ ■ ■

With our ocean and atmosphere backgrounds combined, we like to say we have the world covered. But it's not just us. We

NOTE FROM THE AUTHORS

have an extensive network of wonderful colleagues and friends within similar and related fields. Together, and with their input, in the following pages we provide answers to questions about some of our favorite topics.

Please note that our intent is not to create a comprehensive guide on each topic but rather to provide short, easy-to-understand answers to common, interesting, and sometimes oddball questions. We encourage you to explore the topics covered in more detail and provide sources for additional information at the back of the book. Happy reading!

MEGALODONS, MERMAIDS, AND CLIMATE CHANGE

Chapter One

THE DEEP VAST SEA

MOST OF THE DEEP OCEAN remains unexplored. Very little has been observed or mapped in any detail. Even the experts have a myriad of unanswered questions. It's ironic, as the ocean covers nearly three-quarters of our planet and, along with the atmosphere, is what makes Earth livable or, in some cases, can cause disaster. We also rely on the deep ocean in innumerable ways, including for transportation, food, security, minerals, recreation, and tourism. Here are a few of the questions frequently asked about the deep sea.

AND THE WINNER IS . . .

Drum roll, please. The most frequently asked question about the deep sea is: *Do megalodons still exist?* It is an inquiry often followed by: *What about mermaids?* The no-gratitude award here goes to the Discovery Channel and Animal Planet, which have created the misleading genres of "docufiction" and "mockumentary," including the 2013 Shark Week special *Megalodon: The Monster Shark*; the 2014 sequel, *Megalodon: The New Evidence*; and its predecessors, *Mermaids: The Body Found* (2011) and *Mermaids: The New Evidence* (2013).

With actor scientists, fake investigations, and doctored videos and photographs, the initial megalodon show was so convincing that in a 2013 poll done by the Discovery Channel, 71 percent of the respondents thought megalodons were still alive. Based on the number of people who ask about mermaids, the other shows were also persuasive. Is it because the producers played the dark government cover-up card for conspiracy-loving audiences? Or is it because viewers missed the several-second disclaimer in minuscule print at the end of each program? Or perhaps audiences consider the Discovery Channel and Animal Planet sources of credible science and nature information? In 2010, celebrating its twenty-fifth anniversary, the Discovery Channel identified itself as "the number one nonfiction media company in the world." In any case, we scientists are left trying to correct the record. Here goes.

Megalodons were huge sharks, thought to be the largest sharks ever to inhabit the ocean, and were once very much alive. Most of what we know about them comes from fossilized teeth because for cartilaginous fish, like all sharks, little else is usually preserved. Based on the size of their teeth, scientists estimate adults were bigger than a typical school bus, up to forty-nine to fifty-nine feet in length with a jaw some ten feet wide. Their enormous teeth, up to about seven inches in length, make a great white's three-inch-long teeth seem almost dainty. Megalodon teeth were also serrated, meaning they were meat eaters, probably feasting on other sharks, whales, large fish, seals, and sea lions. Based on the distribution of fossilized teeth, megalodons lived in tropical and subtropical waters throughout much of the world's oceans.

So, megalodons were giant sharks, liked relatively warm ocean temperatures, and needed to eat a lot. That means they probably fed where food is most abundant—the highly productive waters along or near the coast. If they were still feeding in relatively warm near-coastal waters, someone would have seen, caught, or filmed one by now. Nada. Sharks continually replace their teeth, and over a lifetime can produce up to about forty thousand teeth. If megalodons were out there cranking out and dropping tens of thousands of huge teeth, some relatively fresh ones would have been found. Zilch. The only megalodon teeth found are fossilized and millions of years old. Megalodon bite marks have been discovered in the fossilized bones of their prey, but none have been found in the remains or bones of modern animals. Because . . . they are extinct, gonzo, history, aka *not alive today.*

What caused the demise of the world's largest shark? Evidence suggests that in the late Pliocene Epoch (2.6 to 5.3 million years ago), the planet cooled. This meant decreasing ocean temperatures and changes to the food web, which could have reduced the megalodon's dining choices. Warm-water habitats favored by the shark may also have been reduced. And its competitors, such as the great white shark, may have gained the advantage since they required less food. Whatever the cause, megalodons are definitely no longer roaming, lurking, or swimming about in the deep sea.

MERMAIDS?

Let's just put this to rest. The idea of a mermaid or merman is fun, but they are not and never were real. Ariel and Aquaman are fictional characters. Tom Hanks did not rescue an actual mermaid in the 1984 movie *Splash*. Manatees are wonderful sea creatures, and we absolutely love snorkeling with them in Crystal River, Florida, but they are not and never were mermaids, nor were they rotund relatives. Cave paintings of a human/dolphin hybrid are not scientific evidence that mermaids or any such creature once existed (as the author of a book about human/dolphin hybrids contends). In fact, no bones, no fossils, no photos, no videos, no real evidence of mermaids exists (plate 1). As the National Oceanic and Atmospheric Administration (NOAA) posted on their website following Animal Planet's mermaid docu-make-believe, "No evidence of aquatic humanoids has ever been found."

Following a talk, a precocious young man once asked: *If only a small percentage of the ocean has been explored, how do you know mermaids don't exist?* Thanks again, Animal Planet. For fun, let's think about the ocean and the typical image of a mermaid/merman: human up top and scaly fish below.

Marine mammal expert Dr. Shea Steingass has noted several excellent reasons why mermaids as imagined could not exist in the ocean. First off, they would freeze to death. Most of the ocean is cold—really cold. Ocean temperatures drop

Have you even been in the deep sea? You'd freeze your ass off.

dramatically below the surface. At about 650 feet, the average temperature is about 39°F. As Dr. Steingass suggests, to survive in the cold ocean, mermaids would need to be encased in blubber (think of deep-diving elephant seals and whales) and/or extremely hairy (like otters). Extremities like long slender arms would literally freeze off. So, based on temperature, if mermaids were to exist in most of the ocean, they would have to be super hairy and/or sheathed in thick blubber, with stubby to no arms.

Dr. Steingass also considered evolution, citing that fish and humans are far apart on the ancestral tree and our parts just don't match. Then there's reproduction. How would a mermaid's human half happen when reproducing with fish private parts? And what about digestion? A mermaid would consume food like a human and eliminate waste as a fish. Dr. Steingass suggested this scenario would result in a lifelong bout of serious constipation.

Then add pressure to the unholy mix for mermaids. How would mermaids contend with the crushing pressures of the deep sea or changes in pressure when diving from the surface to even average ocean depths (about 12,100 feet)?

Mermaids are not real, were never real, and—like megalodons—are not swimming about out of sight in the deep sea. That's not to suggest there's anything wrong with mermaids if you're a fan. They're mythical creatures that inspire imagination, storytelling, and creativity, and they support a thriving costume industry, mermaid swim classes, and jobs as mermaid impersonators for events, theme parks, restaurants, and parties. We like mermaids but want to be clear about what's real and swimming around in the deep sea and what's not.

WHAT WOULD YOU ASK A DEEP-SEA EXPLORER?

Dr. Susan Humphries is a deep-sea geochemist and scientist emeritus at the Woods Hole Oceanographic Institution (WHOI). She is widely recognized for her groundbreaking work on deep-sea hydrothermal vents and geochemical cycling. She's also done twenty-five dives in the deep-sea submersible *Alvin*. Of all the possible questions to ask about exploring the mysterious deep ocean, one of the most frequent she gets is: *Where is the bathroom on the* Alvin*?* Then again, it's not an inconsequential thing to ponder when submersible dives into the deep sea can last up to ten hours.

Having done more than five thousand dives, the *Alvin*, named after WHOI deep submergence scientist Allyn Vine, is considered the workhorse of deep-sea exploration. Launched in 1964 and refurbished in 2021, the *Alvin* is capable of diving to a depth of more than 21,300 feet, enabling access to about 99 percent of the seafloor. So . . . *Is there a bathroom on the* Alvin*?*

The *Alvin*, like other deep-sea submersibles, is designed to withstand the extreme pressures of the deep ocean. For every thirty-three feet in depth below the sea surface, pressure increases one atmosphere (the weight of the atmosphere at sea level or 14.7 pounds per square inch). At the average ocean depth of 12,100 feet, the pressure is about 370 times greater than at the surface. But in the deepest reaches of the ocean, in the Mariana Trench nearly seven miles down, the pressure is even more extreme, greater than sixteen thousand pounds per square inch; as Dr. Paul Yancey of the Schmidt Ocean Institute suggests, imagine a hundred elephants standing on your head. To dive safely into the deep sea, the *Alvin* has a spherical titanium hull eighty-three inches in diameter with three-inch-thick walls and is surrounded by a titanium frame and syntactic foam. Patrick Hickey, who worked as the *Alvin*'s operations manager and a sub pilot, once equated diving in the *Alvin* to sitting in a phone booth with two of your closest friends, all day long.

Inside the *Alvin*, controls are needed for life support, power, navigation, communication, and scientific observations, photography, and collection. The size, shape, and weight of everything that goes inside must be carefully considered to ensure the *Alvin* has sufficient buoyancy to remain upright and stable during dives. And, in case of emergencies, the submersible carries an extra three-day supply of oxygen. Put all that inside the small titanium sphere, add three people (a pilot and two observers), and there's little room for much else. Occupants must regulate their intake before voyaging into the depths, but if needed, a waste receptacle is available (aka a pee bottle). The bottom line: There is no bathroom on the *Alvin*.

The sea is vertically vast as well as horizontally immense. Submersibles are powered by batteries that are recharged between dives. They're not made to traverse long distances and then dive into the ocean's depths and return. A mother ship is needed to transport a submersible and its crew to areas of interest. It is technically difficult, costly, and logistically challenging to explore the deep vast sea, which is one reason only a small percentage of it has been observed firsthand.

IF SO MUCH OF THE OCEAN IS UNEXPLORED, HOW COME THERE ARE MAPS, CHARTS, AND EVEN 3-D IMAGES OF THE SEAFLOOR?

For most of the ocean, depictions of the seafloor, simply put, are not very good. They are based mainly on satellite altimetry, which estimates bathymetry, or ocean depth, from space. Scientists first determine sea surface height based on the time it takes for a radar pulse from a satellite to bounce off the ocean surface and return. Then, because where more mass occurs (as with an undersea mountain, or seamount), gravity causes the sea surface above to bulge, the depth to the underlying seafloor can be estimated. The problem is that to create a bump in sea level, features need to be at least a mile high and several miles wide. Anything smaller can't be resolved by this method.

In other words, all those charts, maps, and images of the seafloor don't show a lot of what's down there. According to NOAA, there may be as many as a hundred thousand seamounts rising 3,280 feet above the seafloor, yet fewer than one-tenth have been explored. For those traveling in the ocean's depths, this can be a costly and dangerous problem.

In 2021, the nuclear-powered fast attack submarine USS *Connecticut* collided with an uncharted seamount in the South China Sea, resulting in injuries to eleven crew members and tens of millions of dollars in damage. But it wasn't the first submarine to run into an unidentified seamount. In 2005, the USS *San Francisco* hit an uncharted underwater mountain near Guam.

With advances in satellite and ocean exploration technology, tens of thousands of uncharted seamounts have been discovered, and thousands have been mapped in greater detail. Many more, however, remain unfound, unexplored, and uncharted.

WHY IS SO LITTLE OF THE DEEP SEAFLOOR MAPPED IN DETAIL?

Money, time, and logistical challenges. To map the deep seafloor in detail or search for something the size of, say, an airplane or a ship requires specialized instrumentation. And it must be used aboard or towed by ships, submarines, submersibles, or remotely or autonomously operated vehicles. Such operations are costly and logistically challenging, especially in remote areas like the Indian Ocean where Malaysian Airlines flight 370 purportedly went down in 2014. Areas of the seafloor that are well mapped tend to be of high interest or societal value, such as for navigation, resources, transportation, safety, and security.

CAN WE TRACK THINGS UNDERWATER WITH GPS?

When scriptwriters wanted a crack investigative team to track an alien creature under the ocean for a new and exciting television show, they had a problem. The Global Positioning System (GPS) doesn't work undersea. The signal to and from satellites cannot penetrate water. If the show's submerged extraterrestrial was outfitted with a satellite tag, its position could be detected when it surfaced. But in between those locations, the team would have no idea where the creature had been or where it was headed. Story change!

GPS has dramatically improved navigation on land, in the air, and on the ocean's surface. Precise positioning and navigation underwater, especially in the deep sea, remains an enormous challenge and obstacle for exploration, mapping, search and rescue, military operations, and research.

STRANGE LINES ON THE SEAFLOOR?

NOAA's Ocean Exploration Program reports that people often ask if perhaps the strange lines they see on seafloor maps are evidence of a lost civilization. Did they just make the discovery of the century? Sorry, folks, the linear marks or lines commonly seen on the seafloor in bathymetric charts or images are usually artifacts of the mapping process. Seafloor data are collected by various organizations at different resolutions, or level of detail. When combined in a mapping application, it can create gaps or weird, gridlike lines.

ARE THERE UNDISCOVERED SPECIES IN THE DEEP SEA?

Heck yes! We now know that within the deep ocean, once thought devoid of life, organisms not only exist, they can thrive, such as at hydrothermal vents, on whale carcasses, and even in freezing conditions under the ice. As we observe new reaches of the sea and explore other areas in more detail, new species are regularly being discovered. There are tens of thousands, perhaps even millions of species (if you count microbes) yet to be observed and identified. Few of the undiscovered species are likely to be large or charismatic creatures; most will range in size from small to miniscule. Here are just a few of the weird and wonderful deep-sea organisms found relatively recently.

While exploring a seamount some 850 miles west of Hawaii in 2017, a team aboard NOAA's ship *Okeanos Explorer* discovered a seascape of stalked glass sponges they described as Dr. Seuss-like, dubbing it the "Forest of the Weird." One strange sponge found at 6,560 feet reminded scientists so much of a movie extraterrestrial that when it was later identified as a new species and named, it became known as *Advhena magnifica*, or "magnificent alien" (plate 2).

THE DEEP VAST SEA

That same year, northeast of Necker Island in the Hawaiian archipelago at a depth of more than 14,000 feet, an unidentified and ghostly pale octopus caught the attention not only of another team aboard the *Okeanos Explorer* but also of the media, who nicknamed it Casper and made it one of *Time* magazine's "animals of the year" (plate 3a).

During another *Okeanos Explorer* mission some twenty-five miles off the coast of Puerto Rico, crew members using a remotely operated vehicle and high-definition video filmed a colorful new ctenophore, or comb jelly species, at a depth of more than 12,800 feet. Floating like a sparkling hot air balloon underwater (plate 3b), it was later named *Duobrachium sparksae*.

Other recent discoveries in the deep Pacific include new sea stars and sponges, along with novel sea cucumbers, sea urchins, a deep-sea batfish, and a blind eel. As more of the ocean is explored in person and virtually, new and wondrous life forms are and will continue to be discovered. Sadly, even in the deepest reaches of the sea, explorers also find the debris of human society, such as beer bottles and plastics.

WHAT IS THE GREAT PACIFIC GARBAGE PATCH?

In 1997, on their way home from a Los Angeles-to-Hawaii sailing race, Captain Charles Moore and the crew of the *Alguita* decided to take a shortcut through the open North Pacific Ocean. Fishermen and sailors usually avoided the area because of its lack of fish and wind. Captain Moore was shocked by what they found: "It seemed unbelievable, but I never found a clear spot. In the week it took to cross the subtropical high, no matter what time of day I looked, plastic debris was floating everywhere: bottles, bottle caps, wrappers, fragments." The area they crossed soon became known as the Great Pacific Garbage Patch.

The Great Pacific Garbage Patch conjures up an image of an island of human refuse or huge mass of debris floating at the surface. Instead, think of an enormously wide and deep area littered with pieces of plastic ranging from miniscule microplastics to bottle caps, bottles, buoys, buckets, and giant clumps of fishing line. It is a vast soup of plastic stirred by wind, currents, waves, and turbulence. Scientists estimate that the Great Pacific Garbage Patch, located between Hawaii and California, is more than 600,000 square miles in area and contains some 79,000 tons of plastic (possibly about 1.8 trillion pieces).

The North Pacific's garbage patch is not the only such area where plastics accumulate. Additional garbage patches can be found in the ocean's other gyres, or large circulating currents, such as in the South Pacific, the North and South Atlantic, and the Indian Ocean. Floating debris collects at the center of these large-scale rotating currents, created by the distribution of land masses, Earth's rotation, wind, and pressure gradients in the ocean. These flows carry debris toward the center of a gyre, creating a way in but no way out. The distressingly enormous amount of plastic accumulating in the ocean comes from a variety of sources, including fishing, shipping, and offshore aquaculture practices as well as land runoff, river outflows, storm and tsunami debris, and poor waste-disposal practices.

WHY CAN'T WE JUST CLEAN UP THE GREAT PACIFIC GARBAGE PATCH?

Given the extent and distribution of the various-sized plastics in the Great Pacific Garbage Patch, cleanup is not an easy or inexpensive endeavor. To do the job, NOAA estimates it would take sixty-seven ships in service for an entire year. Some efforts are underway to periodically scoop up and remove the plastics, but such activities can also harm marine life clinging to or hiding in the debris.

In addition to removing plastic from the sea, it is critical to reduce the amount entering the ocean. That means using less, reducing production, improving disposal, recycling, and management practices, and additional education. If you see unnecessary plastic packaging at the store, such as plastic-wrapped potatoes or bunches of bananas (why?), don't purchase those products. Use refillable bottles and reusable containers as much as possible. Help others to do the same and to dispose of their waste properly. By reducing our use, we can decrease the demand for plastics and send a message that we want less plastic in our lives and in our oceans.

■ ■ ■

These are just a few of the frequently asked questions we and our colleagues get about the deep vast sea. There's so much more to ask and learn about (maybe a sequel?). We provide sources at the back of the book for further reading and encourage you to explore these topics in more detail.

Chapter Two

DANGEROUS MARINE LIFE

HAVE YOU *ever been bitten by a shark?* For scientists who go to sea or work in the ocean, that's a regularly asked question. For most, the answer is no. The exception might be shark researchers who regularly catch, tag, and study sharks and put their hand close to a shark's mouth. But nobody ever asks if you've been threatened by a leaping smiling dolphin or chased by a furry sea lion. Both are large powerful animals that can be dangerous depending on the situation (personal experience).

Most of the time, it's not what's in the ocean that poses a threat. Human error, poor choices, lack of preparation, or simply bad luck is usually at work when trouble strikes in, on, or under the sea. With a little knowledge and some common sense, the ocean is a safe and wondrous place to work, visit, and enjoy. Here are some questions we and our colleagues are often asked about marine life deemed dangerous.

DO SHARKS ATTACK AND EAT PEOPLE?

Not surprisingly, at the top of the ocean fear-factor list is sharks. With the plethora of books, television shows, posts, and movies portraying them as aggressive, terrifying, and of course people-eating, it's no wonder humans fear sharks. Film industry staple IMDb reports there are more than 180 shark movies, including one of our all-time favorites, *Jaws*, along with six *Sharknadoes*, and the oh-so-realistic multiple-head genre films with two-, three-, four-, five- and even six-headed sharks. Add to the list, movies with mutant sharks, militarized sharks (including Nazi-created flying attack sharks), freshwater invading sharks, sand-swimming sharks, zombie sharks, Jersey-shore and spring-break sharks, a supermarket shark, thawed or reanimated prehistoric sharks, a cocaine shark, and *Doll Shark*, a shark plush toy inhabited by the evil spirit of a killer shark. *Really?*

Sharks are apex predators, at the top of the ocean's predator-prey hierarchy. Because the ocean is opaque, they remain mostly hidden except when at the surface. Given that these big, powerful predators swim about unseen, often in the shadows, some fear of sharks is understandable and healthy. But people are not on the fine-dining or comfort-food menu for sharks. Humans are a poor food choice: skinny, bony, and lacking in blubber or oil. Sharks mostly prefer fish. They also eat crustaceans like lobsters and crabs, squid, octopus, and shellfish, along with other sharks, porpoises, sea turtles, seals, sea lions, and sometimes whales. Some sharks are opportunistic scavengers, eating the already dead if available.

DANGEROUS MARINE LIFE

Every year, billions of people enter the ocean. On average, according to the Florida Museum of Natural History's International Shark Attack File, about seventy people worldwide are bitten by sharks each year. The chances of being killed by a shark are about one in more than four million. More people annually drown or are caught in rip currents. You are more likely to be struck by lightning, die while taking a selfie, or be mortally wounded by a toaster than killed by a shark. Though rare, shark encounters do happen and can result in tragedy. But these are not attacks specifically targeting humans.

Humans are not preferred dining for sharks.

Shark bites typically happen as a result of mistaken identity, in a defensive or territorial response, when sharks are harassed or being fed, or when people fish. Most often, when sharks realize that what they have bitten is unpalatable, people are released or spit out.

Bottom line: Sharks should always be treated with respect. To reduce your risk, swim with a buddy; don't swim at dusk, dawn, around schools of fish, where someone is fishing, or in murky water; and avoid shiny jewelry and excessive splashing.

Now that we've addressed the issue of sharks regularly eating people—they don't—and how best to avoid shark encounters, here are some more frequently asked and oddball questions about sharks.

CAN SHARKS DETECT A DROP OF BLOOD IN THE OCEAN FROM A HUNDRED MILES, OR ONE MILE, AWAY?

Mote Marine Laboratory shark expert Dr. Bob Hueter is often asked some version of this question. Sorry, moviemakers, but the answer is no. Dr. Hueter assures audiences that in the open ocean, one drop of blood would be rapidly diluted. Detection also depends on the shark's location, along with currents, wind, tides, and turbulence. Research suggests that, at best, sharks may be able to detect a drop of blood in a body of water about the size of a small swimming pool. One of us (Ellen) has also been asked if human urine attracts sharks. As far as we can find, there is no scientific evidence indicating sharks are attracted to human urine. And based on personal experience from someone who has lived undersea, spending very, very long hours in a wetsuit, peeing does not attract sharks.

CAN SHARKS BE USED TO WARN OF AN APPROACHING HURRICANE?

Dr. Hueter and his colleagues have shown that blacktip sharks can sense and react to changes in barometric pressure associated with the approach and passing of a tropical cyclone. We'll address tropical cyclone forecasting later in the book, but suffice it to say our system of satellites, reconnaissance aircraft, radar, sophisticated computer models, and instruments is significantly more reliable. And if we were to use sharks, how exactly would that work? Things other than storms can influence their behavior, like a big juicy bait ball, a school of fish, or other sharks. And it's not as though they'd jump up and say, "Hey, a storm's approaching, best to evacuate now before road conditions become too dangerous. What are roads?"

WHY ARE SHARKS DARK ON TOP AND LIGHT ON THE BOTTOM?

This great question came from a middle grader. The color pattern described, dark on top and light underneath, is called countershading. It helps sharks and other sea creatures, such as whales, dolphins, and rays, hide from would-be predators in the ocean's sunlit zone. Remember, the ocean is a hunting ground that is three-dimensional, with predators potentially above and below. If a predator is looking down, being dark on top helps potential prey blend in with the darkness below. If a predator is looking up, being white or light on the bottom helps to conceal prey in the sunlight shining down.

IS IT SAFER TO STAY IN LESS THAN KNEE-DEEP WATER?

Family trips to the Delaware shore were a tradition for Dave. To avoid sharks, people often said not to go deeper than knee-deep. Though many shark incidents do occur in deeper water, especially where people are surfing, snorkeling, or fishing, shark bites occasionally happen in shallow water near shore. Baitfish sometimes lure sharks, such as sand tiger sharks, close to shore. Historically, juveniles have been implicated in non-lethal bites near shore, especially when the water is murky, as less experienced sharks may mistake a swimmer's foot or leg for a fish. Again, to reduce the chances of encountering a shark in deep or shallow water, it is best to not swim alone, stay away from especially murky water, and avoid areas where people are fishing. Also avoid going into the water at dusk, dawn, and around schools of fish. And keep the bling and splashing to a minimum.

WHAT'S UP WITH THE SHAPE OF A HAMMERHEAD'S NOGGIN?

The hammerhead shark's strange bar-shaped head is indeed curious (plate 4). Ellen once got the inquisitive side-eye from a circling school of large hammerheads while diving in the Galapagos. She finds their head shape particularly fascinating—

and the reason for it also cool. The shape of the hammerhead's head enhances its superpower senses. Using mucus-filled pores on their snouts, called the ampullae of Lorenzini, sharks can sense weak electrical fields like those associated with heartbeats or muscle movements. Since the hammerhead's snout is wide, it has an expanded electrical detector at the front. The shark uses it to hunt its favorite food, stingrays, which often lie buried in the sand. Having an eye on each side of its bar-shaped head is also the shark version of a rearview mirror.

MORE THINGS ABOUT SHARKS INQUIRING MINDS WANT TO KNOW.

Two experts, Dr. Gene Helfman and George H. Burgess, wrote an awesome book, *Sharks: An Animal Answer Guide*, to answer questions about sharks. Here are a few of our favorites, slightly modified, with abbreviated answers.

Can sharks see colors, like yum-yum yellow? No. Because of the configuration of their eyes, sharks cannot see colors; they are essentially color-blind. But they can detect brightness and contrast. So yellow may stand out, not because of its hue but because of the brightness or contrast with the surrounding water.

Do sharks talk (or growl)? The undersea world is too quiet for some moviemakers, so they like to add sound when sharks are involved. But no, sharks do not talk, growl, or grunt. They are silent, graceful predators.

Do sharks sleep? A few sharks, including nurse, whitetip, and wobbegong species, can lie still on the bottom. To obtain oxygen while at rest, they use their mouths to pump water over their gills. Other sharks must swim constantly with their mouth open to push water over their gills. Some sharks swim open-mouthed but can also switch periodically to pumping water over their gills. Sharks appear to have active and resting periods, but they do not sleep like humans.

Do nurse sharks bite? Resting on the seafloor or in small caves, nurse sharks appear docile. This has encouraged some snorkelers or divers, playing with less than a full deck, to pull their tail or pet them. Don't do it. As Helfman and Burgess explain, "nurse sharks are one of the few sharks flexible enough to put their tail in their mouths." Oh, and they have teeth. In 2006, one diver went even further and decided to kiss a nurse shark. The return smooch resulted in a hospital stay. Though they have small teeth, nurse sharks reportedly bite and hang on, reluctant to release their grip. So, yes, they bite. Do not—we repeat, do not—pull their tail, pet one, or pucker up.

Do shark repellents work? Yes and no. Research suggests it depends on the shark or sharks, the circumstances, and the repellent. Curious or leisurely feeding sharks may sometimes be discouraged by repellents, but actively feeding and hungry sharks typically ignore electrical pulses, bubble curtains, sonic guns, chemicals or dyes, specific wetsuit colors—essentially all repellents.

As top predators, sharks play an important role in the ocean ecosystem. They keep prey species in check, weed out weak or sick individuals, provide a source of food for other top predators, and are part of the ocean's overall nutrient, carbon, and food webs. If anyone should be afraid, it's the sharks. People kill millions of sharks each year. International experts identify about one-third of all shark species as threatened with extinction. Overfishing and incidental killing of sharks in fishing operations (bycatch) are the main causes of shark decline. Destructive fishing practices such as shark finning (slicing off the fins and tossing back the body) and the decline of nursery habitats are also problematic. But fishing for some species of sharks, such as the spiny dogfish in the United States, is regulated and considered sustainable.

ARE BARRACUDA DANGEROUS?

When a large, silvery, torpedo-shaped fish with a long, pointy snout full of sharp teeth follows close at your fins, it's unnerving. Barracuda do this frequently, but they aren't sizing you up for a meal. They are curious, territorial fish. Attacks by barracuda are exceedingly rare. A few do happen, mostly when people are fishing (do not fight a barracuda or a shark for a fish on a line or spear) and occasionally while snorkeling, possibly due to the flash of shiny jewelry that reflects the sun like a fish scale. Again, leave the bling at home.

HOW ABOUT STINGRAYS?

When wildlife advocate and television showman Steve Irwin was tragically killed by a stingray, it perpetuated the idea that stingrays are aggressive and attack people. With some common sense, backed by a little knowledge, most of the time injuries from stingrays can be avoided.

Stingrays are flattened cartilaginous fish with a serrated whiplike tail hosting spines and venom. The "sting" occurs when a person is stuck or stabbed by the tail. This happens most often when someone steps on a stingray resting or hiding in the sand in shallow water. It's a defensive reflex, not an aggressive behavior. To avoid stepping on stingrays, do the "stingray shuffle"—shuffling your feet in the sand to alert them to your presence so they can swim away. If stuck by a stingray barb, use heat to deactivate the toxin, clean the wound, and seek medical attention as needed. If the barb is embedded, it is best removed by medical experts.

Stingrays are not aggressive and do not purposely attack people. They are peaceful creatures whose winglike modified fins propel them gracefully over the seafloor and through the sea. Most forage at the bottom for food, such as invertebrates (crabs, worms, clams, and shrimp) and small fish.

CAN A PERSON BE SWALLOWED BY A WHALE?

While diving for lobster off Provincetown, Massachusetts, in 2021, Michael Packard reported being swallowed by a humpback whale. In 2020, media outlets reported a humpback whale nearly swallowing two kayakers in San Luis Obispo Bay, California. These events beg the question: *Can a whale actually swallow a human adult?* For baleen whales like a humpback, the answer is a definite no. For toothed whales, it's possible, though extremely unlikely and only for one species, the sperm whale.

Humpback whales feed on plankton, mostly small crustaceans, especially energy-dense krill, and small fish. After engulfing a large mouthful of water and food, these whales use their tongue to force water out through their baleen, which are keratin plates similar in composition to our fingernails or hair. The plankton and small fish that don't go back out through the baleen are swallowed. If a human were mistakenly engulfed by a humpback, they would definitely not make it through the baleen. But there's a bigger (actually smaller) issue—the whale's throat or esophagus. It's only about the size of a fist and, at best, can be stretched to maybe fifteen inches in diameter. So, no, it is not possible for a humpback whale to swallow a human adult.

Only one whale is thought to have a sufficiently large throat to swallow a person—the sperm whale. These toothed, deep-diving whales can grow up to sixty feet in length and feed principally on squid, sharks, and fish. Given that a colossal squid, which can reach forty-six feet in length, was found in a sperm whale's stomach, they can eat really big things. They are also thought to feed on giant squid (different and bigger than colossal squid), swallowing them whole. So, the science suggests a sperm whale could physically swallow a person. But sperm whales do not feed like the engulfing-prey-at-the-surface humpbacks. Instead, sperm whales dive thousands of feet to catch prey. So, it is extremely unlikely that while feeding a sperm whale would (1) encounter a person and (2) mistake them for a deep-sea treat.

ARE ORCAS REALLY KILLER (OF PEOPLE) WHALES?

Seeing orcas hunt larger whales, ancient sailors called them whale killers. It's easy to see how the name got turned around and stuck for a top ocean predator that is smart, fast, powerful, and hunts in packs. But there are no documented human fatalities caused by orcas in the wild, and attacks on people are exceptionally rare. The few reported incidents in the wild are considered cases of mistaken identity, including two encounters with surfers who may have resembled seals. It's unclear why orcas have recently been attacking boats off the coasts of Spain and Portugal. Scientists suggest it may be a form of play or social behavior rather than aggression or "revenge," as some people claim.

Orcas are actually the largest of the small, toothed whales also known as dolphins. They are found throughout the world's oceans but tend to concentrate in colder, more productive areas, where food is more plentiful. Some migrate seasonally, while others are year-round residents in one place. The wide-ranging diet of killer whales includes fish, seals, sea birds, squid, sea turtles, octopuses, sea otters, and other whales. Scientists have even found the remains of deer, moose, and pigs in their stomachs. Killer whales, aka orcas, can be curious, playful, and wondrous animals to behold. Like any apex predator, they should be treated with respect and caution. But killer whales do not prey on, hunt, or regularly attack humans.

Caution is advised when in or on the water near any type of whale. Huge and powerful, one slap of their tail or a fin can cause serious injury. Breaching whales have been known to land on and damage boats. For both humans and whales, it is mutually beneficial to keep encounters at a safe distance; in some places, it's the law. Like sharks, whales have more to fear from humans than the other way around. Entanglements in fishing line, ship strikes, pollution, and overfishing threaten many species (see information on whale strandings in chapter 4).

DO JELLYFISH AND PORTUGUESE MAN-OF-WARS PURPOSELY STING PEOPLE?

Wrong place, wrong time—that's the deal with the majority of jellyfish stings, as most of these creatures simply drift with the wind, waves, and currents. We are asked a lot of questions about jellyfish, especially one specific inquiry. To do jellyfish justice, we decided to give them their own chapter, which comes up next. But first . . . what about the Portuguese man-of-war?

Portuguese man-of-wars resemble jellyfish but are not true jellyfish. They are colonial creatures called siphonophores. The man-of-war is a collective made up of individual polyps or zooids, each with a specialized task such as flotation, feeding, or reproduction. They are most recognizable by their blue, purple, or pinkish float, or balloon, at the surface and long, blue tentacles. Portuguese man-of-wars also drift with the wind, waves, and currents, often landing on beaches. Their tentacles, like those of a jellyfish, are filled with stinging cells (nematocysts), and they pack a powerful punch. Stay out of the water if man-of-wars have been sighted. Beware of them on the beach as well, because even then their nematocysts can fire and cause a burning sting.

If stinging jellyfish or Portuguese man-of-wars have been reported or sighted in the water, it is best to remain land-based.

LOOK BUT DON'T TOUCH

There are just a few other things in the ocean that can sting, bite, or stick you. But they are relatively easy to avoid. In coral reefs, fire coral, fire sponge, and fireworms impart a nasty sting. None of these creatures can chase you. Coral and sponge are literally stuck in place, and small fireworms wiggle around on top of or in the nooks and crannies of a reef. To avoid stings, just follow the motto: Look but don't touch. The same goes for things that can stick you with one or more toxin-laced spines, such as black spiny sea urchins, scorpionfish, and stonefish. No touch, no problem.

ARE SEA SNAKES AGGRESSIVE?

The image of a sea snake wrapped around a scuba diver's leg is enough to keep some people out of the ocean forever. Fear not. Research suggests these sea creatures are not on the hunt but rather just looking for love or trying to avoid aggressive suitors. There are reportedly more than sixty species of sea snakes, all of them found in the Indian and Pacific Oceans. No sea snakes are found in the Atlantic Ocean or the Caribbean Sea. Propelled by a distinctly flat, oarlike tail, sea snakes are reptiles and must surface to breathe. However, some species can dive to eight hundred feet and spend up to eight hours underwater between breaths. They also need to drink freshwater, which can be found in shoreside puddles or at the sea surface when it rains. Their need for warm ocean temperatures and freshwater, along with the pattern of modern ocean currents and past landmass

Sea snakes sometimes cuddle up, looking for love.

changes, are thought to have prevented the spread of sea snakes into the Atlantic and the Caribbean.

Sea snakes tend to be relatively docile, but they are highly venomous. They feed principally on fish and eels. Bites are extremely rare and occur most often when fishermen remove them from nets or during accidental encounters. While diving in Fiji, Ellen saw kids playing with sea snakes, and a dive master pulled one out of the rocks by hand. Nope, we still don't want one wrapping around a leg even if they are just looking for love. As with any wild animal, especially one with venom, it is best to treat all sea snakes with healthy respect and a good dose of caution.

■ ■ ■

We provide our sources at the back of the book and encourage you to explore these topics in more detail.

Chapter Three

JELLYFISH

JELLYFISH ARE remarkable creatures that have been around for hundreds of millions of years. They have survived the ravages of planetary change and continue to flourish in the modern ocean. Some reports suggest jellyfish populations are on the rise. Before we get to why that might be, we'll start with a jellyfish question so popular it made it onto a hit television show.

IF I GET STUNG BY A JELLYFISH, SHOULD I OR SOMEONE ELSE PEE ON IT?

No! You should not pee on a jellyfish sting or have someone else do it. Not only is it gross, but peeing on a jellyfish sting could make it worse. *So, no, don't do it.*

The idea that you should pee on a jellyfish sting is an old wives' tale or urban legend—something that is widely believed but not true. In an episode of the popular television sitcom

JELLYFISH

When it comes to jellyfish stings . . . the answer is NO!!!!

Friends, one of the characters was stung by a jellyfish and someone suggested peeing on it. This brought a whole new level of visibility to the idea, and it continues to pop up in all sorts of popular media. So, let's set the record straight with a little more information about jellyfish and why this is a such a truly bad idea.

Jellyfish are essentially big bags of watery goo with no brain, bones, blood, or heart. Amazingly, they eat, float, smell, detect light, and reproduce with just thin flimsy bodies, a simple nervous system, and a filling of gelatinous (Jello-like) water-based slime, and a few sort of swim (box jellies like the sea wasp). They also have one all-purpose orifice. There's no "in one end and out the other." Yup, they poop and eat in the same place. Okay, there are definite drawbacks to being a jellyfish.

Jellyfish are all carnivorous; they feed mainly on the sea's small drifting animals (zooplankton), small fishes, and other jellies. The key to their hunting success is their gooey invisibility cloak and numerous tentacles with venom-filled stinging cells. Transparency has dual advantages, allowing them to drift up to prey unseen and to hide from would-be predators. But the real gamechanger are their stinging cells, each of which contains a retracted barb, like a notched arrow ready to shoot. When the stinging cells are activated, the barb is released and delivers venom. The strength or toxicity of the venom depends on the type or species of jellyfish. The venom in the sting of some jellyfish is weak, in others it is much stronger, and in a few cases, it is potentially deadly.

WHY DOESN'T PEE WORK?

Two things can cause a jellyfish's stinging cells to release their barbs and venom: (1) contact with skin and (2) a change in chemistry, such as pH (a measure of how acidic or basic a water-based fluid is). Experts do not recommend rinsing a sting with freshwater because the change in the pH from saltwater to freshwater could make more stinging cells release their barbs. Like freshwater, human urine does not have the same chemistry (pH) as seawater, so it could make things worse. Pee is also not as acidic as vinegar (which is recommended); it is the acidity of vinegar that un-notches the arrow or deactivates a jellyfish's stinging cells.

WHAT SHOULD YOU DO IF STUNG BY A JELLYFISH?

If you are stung by a jellyfish, medical experts suggest (1) rinsing with vinegar to deactivate or turn off any stinging cells attached to the skin; (2) removing any tentacles stuck to the skin with a tweezer, rinsing with seawater, or if necessary, scraping the skin with something like a credit card, and (3) applying heat to deactivate the venom.

If other problems result from the sting, such as a severe rash, difficulty breathing, chest pain, or muscle aches, additional medical attention should be sought (see a doctor or go to the hospital). For jellyfish with strong venom, such as the Indo-Pacific sea wasp, additional medical attention should be sought immediately; an antivenom may be required.

If you're planning a trip to the ocean and are worried about jellyfish, check with the local lifeguards to see if jellyfish are present and stay out of the water if they are sighted. You can also bring along some vinegar and be sure to know what to do if stung.

DO JELLYFISH ATTACK PEOPLE?

Again, mostly no. The majority of jellyfish are simple drifters, voyaging the seas driven by waves, wind, and currents. If you happen to be in their path, you may get stung. Jellyfish do not target humans as food. There are however some species of jellyfish, the cuba-medusa, that may have a poor form of vision that allows them to see objects and swim toward them. These jellyfish may pulse toward someone in the water, but they cannot differentiate a person from something like a school of small fish.

AS SUCH SIMPLE CREATURES, HOW HAVE JELLYFISH THRIVED FOR SO LONG?

Jellyfish are incredibly energy-efficient organisms with few needs. Most drift about the sea literally waiting to run into prey. No energy is needed for propulsion. Their transparent, gelatinous body is low maintenance and provides a dual-purpose cloak of invisibility. The goo or mucus that makes up most of their body is 95 percent seawater. Compared to more active and complex creatures, jellyfish require less food to survive and grow. They can also live in a wide variety of conditions. Another reason they've been around so long and sometimes occur in great numbers is their efficient and fruitful reproductive strategy.

Jellyfish have two forms during their reproductive cycle. One form is a flowerish polyplike creature that attaches to the seafloor and can lie dormant. When conditions are just right, their other form, a tiny spaceship-like baby jelly or larva, is released, often in abundance. This can create gobs of jellyfish, also known as a bloom or swarm.

ARE JELLYFISH POPULATIONS INCREASING?

Observations from around the world suggest jellyfish populations are indeed increasing. Several factors may explain the rise of the armed and gooey. The world's jellyfish eaters, such as sea turtles and some fish, have been reduced by overfishing and nontarget mortality in fishing (bycatch), along with the impacts of marine debris and pollution. Climate change is also at play. In a warming world, jellyfish have an advantage because many can thrive in high seawater temperatures and low-oxygen conditions.

WHY ARE NONNATIVE SPECIES OF JELLYFISH FOUND WHERE I LIVE?

Jellyfish are excellent invaders. For example, the Australian spotted or white spotted jellyfish is now found in the Gulf of Mexico and along the shores of North and South Carolina. Thick swarms of warm-water mauve stingers have invaded the waters off Northern Ireland and plagued the shores of the Mediterranean and the Adriatic. In the Black Sea, the American comb jelly is an unwanted and destructive visitor. By attaching to ships' hulls or as larvae in ballast tanks, jellyfish species can travel the world undetected and adapt to a wide range of conditions. When they invade, nonnative jellyfish may outcompete native species and alter the ecological balance. They can also affect fisheries and fish farms, clog intake pipes for power plants, disrupt tourism and recreation, and pose a risk to swimmers. In Japan, a fishing net full of giant Nomura jellyfish even caused a boat to sink.

IS THERE A WAY TO REDUCE JELLYFISH POPULATIONS?

Yes, and it's growing in popularity. Have you ever had jellyfish noodles, or as we like to call them, joodles (plate 5)? In

Asia, jellyfish have long been on the menu, and today they are becoming more popular in other areas of the world. In some places, jellyfish are even served alive. We do not recommend this version of jellyfish dining for many squiggly reasons. But speaking from personal experience, when processed, cooked, and well flavored, joodles are a tasty, crunchy, and surprisingly nonslimy treat!

For ages, jellyfish have been a part of the natural world, part of the complex and interconnected ocean/atmosphere system we call Earth. Jellyfish run-ins can happen and may happen more frequently because of climate change, overfishing, and the accidental introduction of nonnative species. Many jellyfish are harmless or have a weak sting, and most can be avoided. But no matter what, if you get stung by a jellyfish, for goodness' sake, *do not pee on it.*

■ ■ ■

We provide our sources and references in the back of the book for additional information.

Chapter Four

OTHER SEA CREATURES

FOLLOWING TALKS about the ocean, kids are sure to ask: What is your favorite sea creature? That's like asking who's your favorite child. There are simply too many charismatic and fascinating organisms in the ocean to choose from. When kids are asked what their favorite sea creature is, the results are often as expected and sometimes surprising: sharks, whales, dolphins, sea turtles . . . and then . . . jellyfish or even nudibranch. Huh?

Because marine organisms have evolved a long list of interesting ways to adapt to and live in the ocean and there are "many fish in the sea," this chapter could go on and on (or again, part of a sequel?). Here are a few more regularly asked questions about marine life.

WHAT IS SARGASSUM?

Sargassum is a type of brown algae or seaweed with more than three hundred species. The Sargasso Sea, a region in the

North Atlantic, is named after the historic abundance of two free-floating species: *Sargassum natans* and *Sargassum fluitans*. In these species, small, gas-filled, beadlike bladders keep the algae afloat at the sea surface. In the open ocean, clumps, or rafts, of sargassum are an important habitat and food source for marine life.

When it comes to sargassum, however, too much of a good thing can cause serious problems. News headlines have noted the recent abundance, some would say overabundance, of sargassum in the tropical Atlantic. Carried by wind and currents, thousands of tons have drifted across the ocean and piled up on beaches in the Caribbean, West Africa, and Florida, disrupting fishing, recreation, and tourism industries, and in some cases posing a risk to public health. Data suggest the proliferation of sargassum in the region is due to warming seawater temperatures and excess nutrients flowing into the ocean that act like fertilizer. Changing wind patterns may also play a role in the recent proliferation and influx of sargassum.

WHAT CAUSES A HARMFUL ALGAE BLOOM?

Algae are an important component of the ocean ecosystem. But again, sometimes too much of a good thing can be harmful. Algae require sunlight, nutrients, and carbon dioxide to photosynthesize and grow. An overabundance of nutrients, usually nitrogen and phosphorous, can enter the sea from rivers and land runoff carrying fertilizer, sewage, or animal wastes. If an excess of nutrients is available, algae can grow rapidly and in great abundance, creating an algae bloom. Such blooms can block sunlight for photosynthesis and reduce oxygen production. The decomposition of algae that fall to the seafloor uses up oxygen. If waters are poorly mixed or stagnant, low-oxygen, or hypoxic, conditions may develop, leading to fish kills and mortality of other marine life.

Some algae contain toxins, such as the species responsible for red tides. In these instances, an algal bloom can also become harmful because the toxins are consumed or released and lifted into the air (aerosolized) by wave action, affecting marine organisms and sea birds. In combination with an onshore breeze, aerosolized toxins can also cause respiratory distress in beachgoers and for particularly vulnerable individuals, an unwanted trip to the emergency room.

Harmful algal blooms are increasing worldwide because of excess nutrients entering the sea (pollution) and warming seawater temperatures.

HOW DOES ECHOLOCATION IN DOLPHINS WORK?

The makers of sonar (sound navigation and ranging) systems wish we had a more detailed answer to this question. Here's some of what we know. Dolphins and other toothed whales use echolocation to navigate and find prey, especially in murky or deep waters. Liplike flaps in their head, near the blowhole, create vibrations that are emitted as high-frequency bursts of sound or clicks. The shape of a dolphin's or whale's skull and its melon, a large fatty internal structure within the head, helps to transmit the sound in seawater. The pulses of sound bounce off objects, including prey or obstacles, and are then picked up through the jawbone and transmitted to the inner ears. How dolphins and other toothed whales create a 3-D acoustic picture from the return echoes remains a mystery.

An arctic relative of the dolphin may be the sea's echolocation expert. Because the beluga whale's cervical vertebrae are unfused, its neck is super flexible. It can not only nearly do the exorcist 360-degree head swivel but also change the shape of its melon to better direct echolocation sound waves. By bouncing sound off ice floes, the beluga may even be able to "see" around corners. Now that is a cool science superpower.

DO WHALES AND DOLPHINS SLEEP?

Yes, whales and dolphins sleep, but not in the same way humans do. They literally keep one eye open. During sleep, or their restful state, whales and dolphins shut down half their brain and close the opposite eye. This way, while controlling bodily functions such as breathing, they can also stay on the lookout for danger. When sleeping, dolphins and whales may rest on the surface, swim slowly beside one another, or hang vertically. In captivity, dolphins have also been observed to sleep/rest on the bottom and then make repeated visits to the surface for air.

For dolphin and whale mothers, sleeping on the job is a requirement. Newborns don't have enough body fat or blubber to float. For several weeks, including their sleeping hours, mothers must constantly swim to keep their babies at the surface to breathe. How much time whales and dolphins sleep and exactly how they do it varies among species.

OTHER SEA CREATURES

WHY DO WHALES STRAND?

Years ago in the Bahamas, a pilot whale stranded on a small remote island. To get help, a local man called a marine laboratory to the south in the Exuma Islands. Ellen and her staff responded. Fortunately, as the tide rose, they were able to guide the whale off the beach into the water. It took a few hours of careful slow-boat and snorkel wrangling to then guide the whale to deeper water. Happily, it soon dove and quickly disappeared. Why it stranded, we don't know. The configuration of the islands, with their shallow water bays and shoreline, may have confused the whale's echolocation. As with other strandings, it was difficult to tell without further investigation and information.

Whale and dolphin strandings are difficult to witness. Sometimes the animals can be helped back to sea or rescued and rehabilitated, but often large numbers don't survive. Whales and dolphins are built for a life of swimming, diving, and floating in the sea. When stranded, they are literally crushed by their own weight. Research suggests that strandings can occur for a variety of reasons, including injury, infection, starvation, disease, pollution, social factors, and navigational errors. Sometimes mass strandings occur, particularly in species that live in close-knit, socially connected groups.

In the northeastern United States, there has been a series of recent whale mortalities. Some people have blamed a growing offshore wind-farm industry, but necropsies performed on the whale carcasses indicate ship strikes as the likely cause. Other contributing factors could be the ingestion of plastic pollution and climate change, which can cause behavioral change in prey populations, meaning the whales' food is going elsewhere. There is no evidence that wind farms are to blame.

HOW DOES AN OCTOPUS CHANGE COLOR?

While trying to get an octopus out of an intake pipe for a desalination system in the Bahamas, Ellen was involved in an undersea tug-of-war. It was a short battle, with an eight-armed victor. Octopuses are amazing creatures: intelligent, very strong, able to shape shift, squeeze through small spaces, and they can rapidly change color to camouflage within their surroundings. Dr. Roger Hanlon of the Marine Biological Laboratory in Woods Hole, Massachusetts, has been studying octopuses and their relatives, the cuttlefish, for decades. Even he has been surprised by their incredible ability to hide in plain sight. Within seconds, an octopus can change color, texture, and shape to match the seafloor or even mimic the appearance of other creatures. Dr. Hanlon's research has helped to unveil much about the octopus's "smart skin."

An octopus's skin has thousands of small color organs (chromatophores), like tiny muscular pigment-containing elastic sacs. With contractions of individual or groups of chromatophores, an octopus can create an amazing assortment of hues and designs—stripes, polka dots, or waves of color. Initially, scientists were puzzled by the rainbow of colors they can produce because the chromatophores only contain red, orange, yellow, black, and brown pigments. The answer: They also have reflecting cells in their skin that act like a mirror or prism.

Dr. Hanlon's team discovered that octopuses use their relatively large brain and complex eyes to detect brightness and patterns in their surroundings, which they can replicate. A bigger surprise came when they found that most octopuses and their relatives are color-blind. Huh? How do they match color? Research has now shown that octopuses have color-sensing capabilities in their skin. It's not completely understood but imagine the possibilities if we could create materials that sensed color, pattern, and brightness and rapidly replicated them. Can you say "invisibility cloak" and "instant camouflage"?

HOW LONG CAN SEA TURTLES STAY UNDERWATER?

Like other reptiles, sea turtles are air breathers. They periodically swim to the sea surface, take a gulp of air, and then submerge. But incredibly, they can also sleep on the seafloor for up to about five hours. Sea turtles can slow their heartbeat while sleeping, thus using less oxygen. They also have another trick up their . . . not sleeve. Using osmosis through their cloaca, sea turtles can obtain oxygen from seawater. In other words, while sleeping they can breathe through their butts. Neat trick.

HOW DO SEA TURTLES NAVIGATE THE OPEN OCEAN?

For years, Dr. Ken Lohman and his team at the University of North Carolina, Chapel Hill, have been investigating the travels of sea turtles. Through tracking and experimental studies, they've discovered a lot about how sea turtles navigate the ocean. In one series of experiments, they used harnesses to hang variously sized sea turtles in an aboveground pool (plate 6). They then changed the conditions in the pool and recorded how the sea turtles reacted. Here's some of what they've learned.

After digging their way out of a nest on a beach at night, baby sea turtles use the light of the horizon to find the ocean. But once they enter the surf zone, sea turtles orient themselves perpendicular to the waves to swim toward deeper water. After that, it's all about Earth's magnetic field.

When the magnetic field was altered around the pool, the sea turtles immediately changed their swimming direction. For sea turtles, Earth's magnetic field is their GPS or road map. They use magnetic fields to navigate the ocean and to return to the beach where they were born. But the researchers also found that if beaches have a similar magnetic field, for instance on opposite coasts of Florida at the same latitude, sea turtles may mistakenly nest at a different beach.

Another fascinating fact about sea turtles is that the temperature of the nest determines gender. Cooler temperatures beget males, and warmer temperatures produce females. Deeper in a nest there might be more males, while closer to the surface, females. Because of climate change, in some areas more female sea turtles are being born, and there's concern that too few males may lead to reduced populations and less genetic diversity.

WHY DO MANTA AND EAGLE RAYS JUMP OUT OF THE WATER?

Manta and eagle rays often leap into the air and do the sea-creature equivalent of a high-flying somersault before belly flopping back into the sea. Why in the world would they expend the energy to do this? We're not quite sure, but several theories exist. It might be to remove parasites, to elude predators, or as a form of communication or social behavior. Then again, maybe they are simply playing.

WHY DO FLYING FISH . . . FLY?

Flying fish are like small gossamer-winged missiles that leap out of the ocean and glide atop the waves. Their outstretched fins act like a pair of stiff wings, and their tail is used as a rudder. While in the air beside a motoring ferry, flying fish were once clocked gliding at an impressive twenty miles per hour. They also have a penchant for jumping out of waves and onto vessels, entering through portholes or smacking into unsuspecting deckhands (personal experience). Flying fish fly to avoid predators or when disrupted by things such as ships or boats.

CAN SEA STARS REALLY GROW REPLACEMENT ARMS?

Formerly known as starfish, sea stars are astonishing creatures. And yes, if they lose one arm, they can grow another. But wait, that's not all. If an arm is trapped in the jaws of a predator, sea stars can self-amputate the arm and grow a replacement. Sea stars have duplicate internal organs in their arms and blood that is essentially slightly modified seawater. Not only can they regenerate an arm, but if an arm breaks off with part of the body or central disk, they can regrow a body. Whaat? Oh, and one more cool sea star fact: Sea stars can exude their stomach out of their body to feed on shellfish they've pulled open—the ultimate in dining out.

WHAT'S THE DIFFERENCE BETWEEN A SEAL AND A SEA LION?

An easy way to distinguish a seal from a sea lion is to ask this question: Is it waddling on flippers or doing the dance move known as the worm, wriggling forward on its belly? The former is a sea lion, the latter a seal.

Sea lions have external ear flaps and large, long front flippers. While swimming, they use their front flippers as paddles and their rear flippers to steer like a rudder. On land, sea lions walk or waddle on their flippers. They also tend to "bark" loudly and are often found together resting on beaches, rocks, or sometimes jetties, piers, and even boats. They might look cuddly, but sea lions also have large teeth, horrible fish breath, and may bite when threatened or harassed.

Seals have no external ear flaps, have smaller front flippers, and propel themselves through the water by moving their rear flippers back and forth (like a fish). They tend to be quieter than sea lions, and on land they move about on their bellies, kinda like the worm dance.

WHAT'S THE UGLIEST SEA CREATURE?

For this list, young audiences typically nominate the bloblike blob fish, the deep-sea large-mouthed dangling-lure angler fish, the long-snouted goblin shark, or the eellike hagfish. But no creature is truly ugly. It's all in the eye of the beholder, even if undersea.

In case you're wondering about sea creatures that glow, bioluminescence is covered in chapter 6.

■ ■ ■

For more information on the fun facts provided here, please see the sources section in the back of the book.

Chapter Five

CORAL REEFS

WE WRITE THIS CHAPTER with mixed emotions. Early in her career, Ellen spent many hours underwater exploring and studying coral reefs in the Caribbean, Florida, and the Bahamas. The beauty, complexity, and wonder of coral reefs were in part what inspired her to become a marine scientist. She recalls mesmerizing undersea scenes with castlelike stands of pillar coral whose white-tipped tentacles waved gently in the sea's flow and a myriad of colorful fish rising and falling in synchrony atop fields of purple finger coral. One of her favorite places was in St. Croix, U.S. Virgin Islands, where snorkelers could drift through a winding forest of staghorn coral patch reefs teeming with fish. On her last trip to St. Croix, those same patch reefs had been reduced to rubble. She also witnessed firsthand the die-off of black spiny sea urchins in the 1980s, the spread of white and black band disease, and coral mortality due to high-temperature stress.

Today, people mourn the loss of coral reefs around the world because of climate change, development, overfishing, pollution, and invasive species—as if it is a new thing. It's not. In many places, coral reefs are but a shadow of what they once were. So, it is with great passion and great sadness that we answer common questions about coral reefs, clear up a few worrisome misunderstandings, and pray that it is not too late to save a little of what's left and spur growth for the future.

Surprisingly, coral reefs, the largest biologically created structures on the planet, are built principally by petite polyps—each less than the size of a familiar chocolate-covered candy. The diminutive, yet mighty coral polyp is essentially a ring of tentacles around a stomach. But by working together, coral polyps build magnificent, hardened reefs that house millions of marine species, are a vital part of the ocean food web, provide an important source of protein for human populations, help protect the shoreline from waves, storms, and tsunamis, and have an estimated worth of trillions of dollars per year.

WHAT IS A CORAL?

By day, most of the world's stony corals appear much like their namesake, rocks. At night, however, they seem to come alive as hundreds of tiny tentacles extend outward to feed on zooplankton. Most corals are colonial, made up of interconnected polyps sitting in small cups in an underlying calcium carbonate (limestone) skeleton. A single colony may be comprised of hundreds to thousands of coral polyps. Depending on the species, coral colonies and their skeletons form distinct shapes such as a mound, anterlike branches, pillars, fingers, a spiraling plate, or a flattened tabletop.

Stony corals are the reef builders. There are also soft and deepwater corals, but these do not produce massive limestone structures. Key to the building success of the stony corals are the symbiotic algae or dinoflagellates living in their tissues, named zooxanthellae (pronounced zoo-zan-thell-ee). Say it with us: zoo-zan-thell-ee. The zooxanthellae typically impart a yellow, green, or brown color to the coral's otherwise transparent tissue. It is a mutually beneficial relationship. Protected within the coral's tissues from would-be diners, the algae use the coral's waste products as nutrients. The coral benefits as well, using oxygen and organic matter produced by the algae through photosynthesis. And since the zooxanthellae use the coral's wastes as nutrients, the coral has an efficient internal garbage disposal. The algae's removal of waste products and input of food to the coral are key to their ability to create a limestone skeleton.

Adult coral polyps secrete calcium carbonate atop their underlying skeleton, essentially jacking themselves up over time and building the rocky framework of a reef. Only the very surface of a coral or reef is alive; below is hard limestone. To learn about corals and how they grow, geologists drill cores from their skeletons (they can regrow over the infilled holes). After drilling, the cores are sliced thin and then x-rayed to reveal the internal layering or banding. Stony corals typically lay down two layers of calcium carbonate each year, one dense and one less dense, which in an x-ray appear as dark and light bands. The light, or less dense, layer indicates faster growth during optimal seasonal conditions. This banding can be used to examine changes in growth over time and determine the age of a coral.

HOW DO CORAL COLONIES GROW OR REPRODUCE?

A colony of polyps can expand or create a new coral in several ways. One means is by budding off a connected new polyp that is genetically identical to the parent—a clone. Corals can also reproduce by spawning, the release of eggs and sperm into the water, which cross-fertilize and form floating embryos that turn into larvae. Some corals release more developed larvae that are ready to attach to the bottom and grow; this is known as brooding. Branching species can also spread through fragmentation. Under suitable conditions, pieces or fragments broken from one colony can grow, attach to the seafloor and form a new colony.

Remarkably, coral spawning on reefs typically occurs nearly the same night or week each year, and somehow it is a synchronized event. Across the reef and at about the same time, corals release their eggs and sperm. It appears like an underwater upside-down snowfall and attracts a host of hungry diners. Though many eggs are eaten by other marine organisms, synchronized release in large numbers improves the chance of fertilization and survival, and leads to increased genetic diversity. Cues to trigger mass spawning include the phase of the moon, water temperature, and tides.

WHAT SPECIES IS A WHITE CORAL?

People sometimes refer to or ask about white coral. The underlying calcium carbonate skeleton of a stony coral is white, but they are not usually referred to as white corals. Perhaps this is because when a reef coral is white, it is typically sick, stressed, or worse—dead. When stressed, stony corals may lose or eject their color-giving zooxanthellae. Their limestone skeleton becomes visible through the transparent tissues, and they appear white or bleached (plate 7).

If the stress is acute or chronic enough, bleaching is fatal. In other cases, a coral may regain its algae partners and remain viable. Because most corals live in an optimum temperature range, prolonged increased or significantly decreased water temperature is the most common cause of bleaching and related mortality.

No, it's not pretty; white coral is either dead or dying.

White coral sold in curio shops and as jewelry are the dead skeletons of once healthy colonies. Imagine putting the skeleton of an endangered or beloved penguin or snow leopard on a table for decoration. Seems unthinkable. While beautiful, much of the coral sold is collected illegally. If you must have one for decoration or in an aquarium, it is best to purchase a coral skeleton reproduction rather than the real thing plucked from, and possibly stolen from, the sea. Also refrain from purchasing red or black coral used in jewelry and carvings; these are deeper-water species that are harvested unsustainably and in a process that destroys important marine habitat.

ARE CORAL AND CORAL REEF GROWTH THE SAME?

No. But this is often misconstrued. Among corals, the speedsters are the branching species, which can grow or extend up to about eight inches a year. That's honking fast for a coral. Mound, finger, and platy corals grow much more slowly, typically less than half an inch per year, depending on conditions. Temperature, light, turbidity, pH, depth, and water quality/chemistry influence coral growth rates.

Reef growth is more complicated. It is essentially the balance between the constructors, including calcifying corals, coralline algae, and sediment-producing calcareous algae, and the destroyers, such as boring mollusk and sponge, grazing sea urchins, and fish. To the complex recipe for reef growth, add factors such as oxygen levels, changes in depth due to sea level, tides, storms, invasive species, influx of sediment due to dredging or development, and more. Estimates of annual coral reef growth range from a rapid 0.78 inches to zero to the negative numbers where growth has stopped and/or degradation dominates.

IS THERE WEATHER UNDER THE SEA ON CORAL REEFS?

Yes! On sunny days with few clouds, a shallow coral reef can be bright and colorful to depths of about sixty feet. Below that, things begin to dim and red light disappears, even on sunny days. As sunlight enters the ocean, short-waved light like green and blue penetrate deeper. Long-waved red light is absorbed more quickly. So, below about sixty feet, without artificial light, everything appears blue-green. At seventy feet, a wound freakishly appears to bleed alienlike blue-green blood (personal experience).

With cloud cover, the sunlight shining into the ocean is diminished and, even at shallow depths, coral reefs appear duller and less colorful. High wind and storms can generate waves that break on a reef, creating dangerous currents and strong surge. A shallow coral reef is no place to be during strong storms, even for corals. Broken branches and mounds can become undersea bowling balls and do significant damage.

HOW MUCH DO CORAL REEFS PROTECT THE SHORELINE?

Mariners have long known the dangers of a coral reef. Running into or onto one is akin to going aground on rocks and often results in being ripped apart and/or sinking. This is because, as noted previously, except for a thin layer of live organisms at the surface, a coral reef is mostly limestone rock.

The hardened structure of a reef and its complexity can also cause waves to slow and break, reducing the energy that strikes shore. But how much the wave energy is decreased depends on the depth, size, and nature of the coral reef and the dynamics of the waves. Coral reef expert Bill Precht confided his frustration at one study that has led to widespread misunderstanding of this issue. Like a game of internet "telephone," it led to erroneous headlines and many organizations using figures that just don't add up.

The study, published in 2014, stated in the abstract that coral reefs "reduced wave energy by an average of 97 percent." The finding was based on data reported in 255 publications, only 10 percent of which directly addressed the issue of wave attenuation by reefs. More specifically, the results suggesting a wave energy reduction of 97 percent involved only thirteen reefs. Coral reefs occur in more than a hundred countries and number, at the very least, in the tens of thousands (depending on your definition of a reef). Thirteen is an insufficient sample size to determine average wave-energy reduction for reefs in general, especially given how much they vary in depth, size, and shape. For Indo-Pacific reefs with a wide reef crest at or near the surface, with high roughness and complexity, a 97 percent reduction in wave energy is reasonable. But for Caribbean reefs that typically have a narrow reef crest near or below the sea surface, this is unrealistic.

Coral reefs provide important protection from waves, storms, and tsunamis, but the extent of that protection depends on ocean conditions and the shape, depth, and topographic complexity of the reef.

IS CLIMATE CHANGE KILLING CORAL REEFS?

Trying to protect coral reefs from climate change is like playing a losing game of whack-a-mole with rapidly rising ocean temperatures, increasing ocean acidity, accelerated sea-level rise, increased susceptibility to disease, and stronger storms. This has never been more apparent than in 2023 as an unprecedented marine heat wave struck much of the ocean and led to the highest average global sea surface temperatures on record. In some areas, warming from El Niño added to the growing crisis for corals. Reefs across the world were bathed in hot tub–like water temperatures for weeks to months. For scientists and many others, what ensued was nothing short of shocking—fields of bleached and dying corals. Sadly, for those corals that did survive the extreme heat of 2023, 2024 is shaping up to be another punch in the gut. By April, the Great Barrier Reef was experiencing its largest mass bleaching event on record. Atlantic ocean temperatures were unusually high and expected to get even higher in the upcoming months in the Caribbean, Bahamas, and off Florida. Even if some corals survive these devastating bleaching events, future prospects are bleak. Reefs are facing not just one or two extreme temperature events but a series that prevent recovery.

Mass coral mortality events are now happening across the world's tropical seas. While the extent of damage in 2023 was unprecedented, it wasn't the first or even second time heat-stressed coral fatalities have occurred. A particularly severe mass mortality occurred in 2015 and 2016. Scientists and conservationists had already begun to take extreme measures to save coral because of stony coral tissue loss disease (SCTLD). First reported in 2014, the disease caused widespread mortality off Florida and in the Caribbean. Notably, corals appear more vulnerable to disease in higher temperatures. To save and protect corals from extreme heat as well as disease, thousands of corals were removed from reefs and placed in tanks with plans to return them when conditions improved. The success of this strategy remains unclear, especially as rising ocean temperatures reach or remain at levels unsuitable for coral growth.

As if scorching ocean temperatures aren't bad enough, corals must also contend with increasing ocean acidity. Increased carbon dioxide concentrations in the atmosphere cause more carbon dioxide to be absorbed in the ocean, lowering its pH (an increase in acidity). With increased acidity, it is more difficult for organisms such as corals to create skeletons or shells of calcium carbonate. Changes in pH also affect the biological functioning of marine creatures, such as respiration, breeding, and growth.

Next up: rising sea level. The melting of land-based ice caps and glaciers, along with the thermal expansion of seawater, creates another problem for coral reefs. If sea level rises too fast for corals and reefs to keep up, the increased depth may not allow zooxanthellae adequate sunlight for photosynthesis.

Another issue for coral reefs: more intense and destructive tropical cyclones. Again, imagine big mounds or thick branches of limestone breaking off a reef and rolling around like a swarm of bowling balls.

As the climate warms, the number of dead zones due to hypoxia or low oxygen levels along our coasts is also rising, and research suggests that oxygen levels in the wider ocean are declining as well. The impacts on coral reefs are unclear, but be assured, it won't be good.

Climate change is decimating coral reefs, and the effects of overfishing, pollution, development, and invasive species are exacerbating the situation. Sorry, but that's reality. Corals and reefs will not disappear entirely or go extinct. A limited amount of coral will probably withstand changing conditions or survive in refugia. And reefs may still exist, but they won't be as we've known them. Surfaces are already and will be further littered with rubble, the dead skeletons of once live coral. Other organisms, such as algae or sponges, will become the dominant players, resulting in less topographic structure and complexity to support and harbor other marine life. Bottom line: Due to climate change, coral reefs are teetering on the edge of catastrophe. And in many areas, the extreme heat of 2023 and 2024 may have shoved them off.

CAN CORAL RESTORATION SAVE REEFS?

Over the past several decades, coral restoration efforts have expanded dramatically, and there's a new urgency to save and protect what's left in the wild. Underwater nurseries and laboratory aquariums are being used to grow and store corals so that they can later be transplanted to areas of damage or coral loss. The fast-growing branching corals were the first to be successfully grown and transplanted onto reefs. Other types of corals are now joining the ranks of nursery-kept and -raised species. Research is also being done on how genetic diversity influences coral survival, how to spawn corals in the laboratory, and the ability of so-called super corals to resist changing conditions.

An important question that must be asked in coral restoration is what constitutes success. Raising and transplanting corals is good, but not enough. The real measure of success is survival, growth, and reproduction in the wild over time. And that has been problematic. Transplants often survive the immediate translocation for up to a year, but in the longer term, survival rates drop dramatically. If the conditions that contributed to the demise of a reef or corals remain or return, transplants typically suffer a similar fate. It appears the extreme marine heat of 2023 has sadly proven this out. Branching corals grown in nurseries and transplanted onto reefs in Florida and the Caribbean suffered staggering losses due to bleaching and mortality. The coral restoration community is now rethinking its strategies. Growing and replanting corals grown from fragments is essentially planting clones of the original coral. If the original is vulnerable to high water temperatures, so are the clones. More reliance may be needed on techniques that increase genetic diversity and/or produce more resilient corals.

Citing the number of corals transplanted on previous projects, coral restoration has been adopted as a mitigation strategy for construction projects that destroy reefs. Hold on. As noted previously, the number of corals grown and/or transplanted is not a metric of success. Survival over time is. Alive, success. Dead, failure. And don't even get us started about the suggestion to saw a coral reef off at its base and transplant the entire thing. Okay, maybe it could be sawed off and possibly moved, at great expense, to a new location, where it could be cemented to the bottom. Maybe. But will a reef, a complex interconnected structure of diverse organisms interacting and dependent on the nature of the underlying substrate and surrounding seawater conditions, survive and grow in a new location? Can you say "crazy town"?

No amount of coral restoration will help if the conditions killing natural corals are not addressed. Climate change, pollution, overfishing, invasive species, and the effects of development cannot be mitigated away.

Coral nurseries and restoration are now part of a growing and important effort to save reefs and preserve corals for the future. Hopefully, improving methods will enable greater survivability in the wild. But such efforts will be futile if more is not done to stop climate change and reduce human-related activities that harm coral reefs. And for all those well-meaning people who suggest things like sunshades, algae-eating crabs, cooling fountains, soothing undersea noise, etc. to save the world's reefs: Even if we could scale up such ideas enough to have a significant and positive impact (doubtful), there would likely be unintended consequences, and results would be limited without fixing the underlying causes of coral demise—in particular, climate change (more in chapter 10).

■ ■ ■

We provide our sources at the back of the book and encourage you to explore and learn more about coral reefs.

Chapter Six

SUPERNATURAL, SUSPICIOUS, OR SCIENCE

THE DUBIOUS, the unexplained, and the mysterious attract attention and lure in audiences. But are strange events and occurrences in the ocean or the atmosphere truly the result of phenomena outside the laws of nature, something beyond our understanding, or a government conspiracy? Or can they be explained by science? Here are some of our favorite questions and explanations related to attention-capturing phenomena in the ocean and the atmosphere.

IS THE BERMUDA TRIANGLE A REAL THING?

Ships disappearing without a trace, flights that mysteriously vanish, and compasses spinning out of control—these are some of the alleged strange goings-on in the area known as the Bermuda Triangle between Miami, Florida; San Juan, Puerto Rico; and Bermuda. Explanations for these events span the supernatural

SUPERNATURAL, SUSPICIOUS, OR SCIENCE

gamut, from alien abductions to ghostly hauntings, and pseudoscience such as huge ship-sinking methane gas bubbles and giant rogue waves. What's fact versus fiction?

Let's look at a few of the facts. Ships and aircraft regularly transit the Bermuda Triangle, but so do hurricanes, tropical storms, and powerful squalls. Periodic high seas, strong currents, thick fog, waterspouts, and turbulence in the region add to the potential for disasters at sea or in the air. Many of the Bermuda Triangle "mysterious" disappearances happened before we had satellites to observe and track hurricanes. Operational hurricane models and computers weren't available to forecast a storm's track or predict its intensity. Ships and aircraft didn't have GPS for precise navigation until the 1990s. And much of the area within the Bermuda Triangle is remote and far from land, so getting help quickly was and still can be difficult, especially at night.

Author and pilot Larry Kusche thoroughly investigated more than fifty supposed ship and aircraft disappearances in the Bermuda Triangle from 1840 to 1973. Through painstaking meteorologic detective work, he discovered that severe weather and high seas could account for many of the "mysterious" air and sea accidents. He found that losses often happened at night, and it was hours before a search and rescue could be initiated. Plenty of time passed for debris, if any, to sink or drift out of the area, accounting for the "missing without a trace." Kusche also uncovered evidence of explosions, human error, mechanical problems, potential piracy, plain old running aground or slipping a mooring, and our favorite, disappearances attributed to the Bermuda Triangle of ships or aircraft that weren't even in the area.

SUPERNATURAL, SUSPICIOUS, OR SCIENCE

The Bermuda Triangle, he concludes, is a "manufactured mystery" furthered by people who (1) avoid fact-checking, (2) add some imaginative speculating, (3) include a few unanswered questions, and (4) use a lot of exclamation points!

But even with modern navigation and advanced technology, the ocean can be perilous. Our personal experiences (Ellen) include nearly capsizing on a small research vessel while crossing the Gulf Stream and again upon entering a notoriously treacherous inlet; going aground late at night during a storm when a boat's anchor dragged; and being on a collision course with a hurricane in the open North Atlantic aboard a 130-foot sailing-school schooner with ten crew and twenty-five undergraduates. On this last occasion, the captain (Phil Sacks) and the chief scientist (Ellen) had to decide whether to cross the path of Hurricane Frances or turn tail but remain in the more dangerous right-hand quadrant of the storm. Captain Phil wisely chose to cross the path of the hurricane, resulting in only minor damage and no serious injuries.

Today, there are far fewer mysterious disappearances at sea, in the Bermuda Triangle or elsewhere, partly because of improved weather forecasting and tracking, navigation, technology, and the emergency position-indicating radio beacon (EPIRB). When activated manually, or automatically if submerged, the EPIRB fixes a location using GPS and emits a distress call to search and rescue forces worldwide. All recreational and commercial vessels should be outfitted with an EPIRB, and one wonders why aircraft regularly transiting over the ocean are not required to have one (e.g., Malaysian Airlines flight 370).

SUPERNATURAL, SUSPICIOUS, OR SCIENCE

For now, there's no solid evidence or data indicating there's anything supernatural or suspicious going on in the Bermuda Triangle.

HOW ELSE CAN YOU EXPLAIN SUDDEN DISAPPEARANCES AT SEA OR IN THE AIR?

Some of the mysterious vanishings in the Bermuda Triangle may have been caused by downdrafts, also known as microbursts. The swift and tragic sinking of two tall ships showcase the extreme danger posed by microbursts. In 1961, six lives were lost when the twin-masted *Albatross* sank 125 miles west of the Dry Tortugas, an incident later depicted in the movie *White Squall*. The *Pride of Baltimore* was next to fall victim to a deadly and sudden downdraft. The ninety-foot schooner, a replica of a nineteenth-century Baltimore clipper and an at-sea goodwill ambassador, had logged more than 150,000 miles over nine years of sailing. Then, on May 14, 1986, some 240 miles north of Puerto Rico, a sudden blast of hurricane-strength winds hit. Knocked down and rapidly taking on water, the ship sank within two minutes. There was no time to radio for help or activate their emergency beacons. Eight crew survived, while tragically the captain and three others were lost. A joint inquiry by the National Transportation Safety Board and the U.S. Coast Guard later concluded that a microburst was to blame.

Former president of the Sea Education Association and tall ship captain Peg Brandon recalls the emotional impact these events had on the tall-ship community. With friends and colleagues lost, it brought new attention to the danger of microbursts, the issue of ship stability, and the need for additional training.

The National Weather Service describes a microburst as a localized column of sinking air (downdraft) within a thunderstorm that is usually less than or equal to 2.5 miles in diameter. In a microburst, winds race downward, strike the surface, and spread out, reaching up to one hundred miles per hour, the equivalent of a Category 2 hurricane or EF-1 tornado.

The danger of microbursts was known before the *Pride of Baltimore* tragedy, but focus had been on aircraft, particularly during takeoff and landing. In 1975, Eastern Airlines flight 66 crashed while attempting to land at New York's JFK International Airport. One of the originators of the tornado severity scale, meteorologist Tetsuya "Ted" Fujita, suspected a microburst was involved. He'd previously theorized the existence of microbursts based on a starburst pattern in the uprooted trees after a tornado in the Midwest. Fujita was part of a team that later used Doppler radar to detect and identify some fifty microbursts. Their work led to the installation of Terminal Doppler Weather Radar (TDWR) systems at airports across the United States and mandatory training for pilots on microbursts and wind shear (a change in wind speed and/or direction with height). It's suspected that before the introduction of these systems, sudden downdrafts may have caused up to twenty major airline accidents and five hundred fatalities.

In 1985, when Delta Airlines flight 191 careened off the runway at Dallas/Fort Worth International Airport, 135 lives were lost; once again a microburst and wind shear were to blame. The accident prompted additional improvements in aviation safety, including the development of wind-shear warning technology and a requirement that all commercial aircraft have wind-shear detection systems onboard.

Sudden downdrafts and dramatic wind shifts associated with thunderstorms can also create extremely hazardous wildfire conditions, as occurred in the 2013 Yarnell Hill Fire, which resulted in the tragic deaths of nineteen Granite Mountain Hotshots.

Given their small size and sudden nature, microbursts are notoriously difficult to forecast or detect. It is one reason the National Weather Service emphasizes the seriousness of their severe thunderstorm warnings. Powerful and deadly, yes. Suspicious or supernatural, no.

WHOA! WAS ATLANTIS FINALLY FOUND... IN THE BAHAMAS?

Within the Bermuda Triangle sits Bimini, Bahamas. Off the island's northwest coast is a site dubbed Bimini Road. Here, rows of large square limestone blocks lie submerged about twenty-three feet below the sea surface. For "true believers," the linear nature of the formation and the shape of the stones are evidence of the long-lost city of Atlantis or perhaps the work of an extraterrestrial construction crew. Some people insist the stones emanate a powerful force field and recommend that to best feel and experience it, a person should be . . . naked. Scientific research by geologists has proven otherwise (not the naked part).

Carbonate sedimentology expert and former U.S. Geological Survey geologist Eugene Shinn was caught up in much of the hoopla about Bimini Road. Asked to investigate, he and colleagues used a specialized underwater drill to core the Bimini Road stones. Laboratory analysis of the cores revealed layers of rounded beach pebbles dipping seaward that could be traced from one stone to the next. From the cores, they also created thin sections, wafer-thin slices of rock that are adhered to microscope slides. Shinn distinctly remembers looking through a microscope and seeing fibrous cement crystals around the pebbles. Rounded pebbles cemented by fibrous crystals in layers that dip seaward is typical of . . . beachrock. Dating techniques also indicated that the stones were less than two to four thousand years old, and no other ancient human or out-of-this-world debris has been found in the area. It's beachrock.

Beachrock forms in the intertidal zone, most prevalently in the tropics and subtropics, where seawater is often supersaturated with calcium carbonate. Repeated wetting of the beach by tides, waves, and spray followed by drying results in the precipitation of a fibrous calcium carbonate crust around the sand grains, which acts like cement. Over time, a hard rock (limestone) forms. Rain, which is slightly acidic, can cause the dissolution of some calcium carbonate sands and when followed by drying results in the precipitation of a crystalline crust and cementation. Years ago, the folks who replenished the sand on Miami Beach learned this lesson the "hard" way. A calcium carbonate–rich sand was chosen to renourish the beach. Some of the grains were made of aragonite, a form of calcium carbonate that dissolves in slightly acidic freshwater. Every time it rained, some of the aragonite dissolved, and when the sand dried, another form of calcium carbonate reprecipitated—creating a cement beach. Oops.

Other processes that may be at play in beachrock formation include the degassing of carbon dioxide from within the sand and microbial activity. Cementation takes place very rapidly, and often the remains of human activity such as bottle caps and other debris are fixed in place during formation.

The Bimini Road stones are beachrock that has been submerged by the rising sea and exposed by erosion of the overlying sediment.

WHAT MAKES THE OCEAN GLOW AT NIGHT?

Our colleagues provided some truly odd, yet wonderful questions they've been asked, but perhaps the winner of the bunch came from marine biologist and bioluminescence expert Dr. Edie Widder, CEO of the Ocean Research and Conservation Association (ORCA): *Was Jesus bioluminescent?* The person asking thought it could explain the (alleged) image of Jesus on the Shroud of Turin. A relative of ours suggested this idea could have been inspired by a glow-in-the-dark Jesus, like the one he had growing up at home. Written with no religious implications, we are very sure Jesus was not bioluminescent, nor was the glow-in-the-dark statue.

All the nifty glow-in-the-dark stuff now available in stores and online, such as necklaces and bracelets, sticky stars for the ceiling, paint, and play slime are also not bioluminescent. To glow, they rely on phosphorescence, in which substances or phosphors are energized by natural light and then glow for a period of time. Two common glow-in-the-dark substances used in toys are strontium aluminate and zinc sulfide. Phosphorescence differs from fluorescence, which occurs when light is absorbed and then almost immediately is reemitted. These two types of luminescence are often confused with bioluminescence, which is the chemical production and emission of light from within a living organism. Two chemicals are generally used by organisms to create light: one, luciferin, produces the light, and the other, luciferase, drives the reaction. Organisms may have organs known as photophores in which this chemical reaction takes place, or they may host symbiotic bacteria that are bioluminescent.

Hoping for the supernatural or a mysterious unknown, a television producer once asked about a strange, glowing, green sphere observed near a boat in the Bermuda Triangle just before the engine suddenly quit. Sorry to disappoint, but it's just science—a bioluminescent jellyfish, that is, which could have clogged the cooling intake of the engine. Scientists now recognize that a large proportion of the organisms in the ocean, up to 90 percent in the deep sea, can produce light. In the ocean, bioluminescence is usually blue-green and occurs in a wide variety of forms, including sparkling, pulsing glows and even vomited or emitted strings or balls of glowing slime (plate 8).

Through observations, including with the use of specialized cameras aboard remotely operated vehicles and submersibles, and through laboratory experiments, scientists have learned a tremendous amount about bioluminescence and its purpose. Take, for example, the stunning experiment in which a squid was placed in a dark tank with a light overhead and a mirror below. When the light was turned on, the squid's image vanished in the mirror below! Sensing the downwelling light, the squid emitted matching light from photophores on its belly. Imagine a predator looking up in the ocean with dim light filtering down. With its bioluminescent superpower turned on, the squid's silhouette would be hidden. Isn't science cool?

The most common form of bioluminescence observed in the sea is a nighttime sparkling or glow in a crashing wave, in a boat's wake, or when a diver waves a hand through the water. Glowing phenomena such as these are typically the result of small dinoflagellates (single-celled algae). Water movement triggers the dinoflagellates to flash, creating a burglar alarm to startle predators or to attract another hunter to target the original threat. Many organisms, including jellyfish, octopuses, crustaceans, squid, and a variety of fish, use bioluminescence not only to scare away would-be diners but also as a decoy, to camouflage, to lure in prey, and for communication.

People sometimes ask if bioluminescence indicates toxicity. Not necessarily. However, it is best to be cautious. Some dinoflagellates or algae contain toxins, but others do not, like *Pyrodinium bahamense*, the amazing algae that create Puerto Rico's dazzling bioluminescent bays.

Bioluminescence is wondrous to behold and important for marine organisms; it is also of interest to the military. In 1953, when the instruments aboard his airplane failed, U.S. Navy pilot Jim Lovell (later aboard the famed *Apollo 13* mission) used the bioluminescent wake of the aircraft carrier USS *Shangri-La* to locate the ship and land safely. Before that, in 1918, during World War I, the bioluminescent wake of a German submarine reportedly gave away its position and led to its sinking by a British ship. Bioluminescence remains of interest to the military in tracking submarines, torpedoes, and ships, along with concerns about keeping secret amphibious landings secret.

Research to better understand, measure, and detect bioluminescence continues today, seeking to learn more about life in the sea, to derive applications in biotechnology, and to foster safer, more effective military operations.

ARE THERE GIANT WHIRLPOOLS IN THE OCEAN?

There are real whirlpools in the ocean, but also imposters: large swirling eddies that from space kind of look like whirlpools. Off the southeast coast of the United States, the warm Gulf Stream flows north before swinging east and then northeast off North Carolina heading toward Europe. Along the way, this strong western boundary current (on the western edge of the North Atlantic Ocean) curves and meanders. Sometimes large swirling rings also known as eddies pinch off from the Gulf Stream's meanders. They appear particularly dramatic in satellite-derived images of sea surface temperature (plate 9). To the north, warm water pinches off into colder water and forms clockwise-rotating warm-core rings. To the south, cold-water masses break off into warmer water and form counterclockwise-rotating cold-core rings. The formation of swirling eddies also occurs in association with the ocean's other strong western boundary currents—the Kuroshio Current flowing from the Philippines northeastward past Taiwan and Japan, and the Agulhas Current, which flows south off the east coast of Africa. Rotating eddies or rings transport heat, nutrients, and marine life, help to mix the ocean, and look super cool from space. They aren't actual whirlpools, but there is one elsewhere in the ocean: the Great Whirl.

Each year around April, the Great Whirl forms off the coast of Somalia. It typically spins clockwise for up to nearly two hundred days, reaches up to 335 miles in width, and can extend to depths greater than six hundred feet. How long this giant whirlpool persists and its timing vary, but research indicates that it generally forms about two months prior to the southwest monsoon season. Its formation and circulation are due to the configuration of the coast, the dynamics of the atmosphere, and ocean influences such as the Somali Current and the annual arrival of slow planetary waves (Rossby waves). Monsoon winds intensify the Great Whirl, and when the winds die down, it dissipates. Complicated, yes; awesome, yes, supernatural, no.

ARE THEY . . . FLYING SAUCERS?

Frequently mistaken for an alien spacecraft or UFO, this atmospheric phenomenon has elicited explanations that include strange mountain magnetism, nonexistent volcanic eruptions, a plane crash, and weather manipulation by a secret or not-so-secret government agency. A stack of flying saucers or floating pancakes are appropriate descriptors, but the formation of lenticular clouds is all about down-to-earth science (plate 10).

SUPERNATURAL, SUSPICIOUS, OR SCIENCE

They're not supernatural or spaceships. Dave loves lenticular clouds.

Lenticular clouds form when strong winds and a stable moist air mass interact with an obstacle, most often a mountain. They typically form parallel to the direction of wind flow atop a mountain or in its lee, the downwind side. Strong winds hit the mountain and are forced upward and around. As the winds rise over the mountain, the air cools and moisture condenses to form a cloud. On the other side of the mountain, the air sinks and dries, cutting off cloud formation. Drier air going around the mountain-top cloud shaves off the sides, creating a saucer-like shape.

Lenticular clouds may form one atop the other with dry air between to create a stacked appearance. They can also create a series of saucerlike clouds over the peaks of atmospheric waves generated by an obstacle in the wind's path; these are known as wave clouds. Lenticular clouds can be fleeting or last for hours and, when they happen at sunset, can produce a seemingly unearthly spectacle. Dave loves clouds, and lenticulars are one of his favorites. But be wary of some "artistic" depictions of lenticular clouds—they're a little *too* amazing to be real.

ROLLING BREAKERS IN THE SKY?

These clouds are sure to make you look up. For the science geeks out there, they are known as Kelvin-Helmholz waves. They form as air rises and condenses to create clouds. Wind blowing across the top of the clouds at faster speeds creates a rolling or breaking wave appearance (plate 11). The atmosphere here may appear like the ocean, but there's no surfing these waves; just look up and enjoy their awesomeness. Dave considers these clouds "atmospheric speed bumps," because flying over them would be a bumpy ride.

WHAT IS FOG?

Need an eerie scene or creepy setting? Call in the fog or mist. Fog machines are the key to an artificially ethereal mood. They rely on a nontoxic chemical or dry ice to create movie magic.

SUPERNATURAL, SUSPICIOUS, OR SCIENCE

Nature has another way of doing it, and the two main actors are air temperature and the dew point.

The dew point is the temperature at which air is saturated with moisture or water vapor. The higher the dew point, the more moisture the air can hold. When the air temperature matches the dew point, condensation occurs and a cloud, fog, or mist forms. *What's the difference?* Mist occurs when fine drops of water become large enough to gently fall out of a cloud. With increasing size, such drops go from mist to drizzle to all-out rain.

Fog is essentially a cloud at the surface and can form in a variety of ways, all of which require cooling and moisture. The thicker the fog, the more moisture involved. Ground or radiation fog occurs when low-lying moist air is cooled to its dew point by the underlying land. In movies this type of fog creates scary scenes, such as a creepy cornfield that kids must venture through. Sea fog, steam fog, or land fog is created when warm moist air moves over a relatively cooler surface. This can also be advection fog, in which light wind moves (or advects) the fog into other areas. Adding moisture to already cold air can also produce fog. If it's extremely cold air, the result may be ice fog. This fog freezes on contact with any solid surface. Bottom line: If there's moisture in the air and the temperature cools to the dew point, voilà—fog.

Fog has impacts beyond looking spooky. In San Francisco, California, fog is the number one reason for flight delays, and many vehicle accidents occur when drivers hit the brakes too late. On the road in fog, be sure to slow and turn on the flashers; it could save lives, yours and others'.

SUPERNATURAL, SUSPICIOUS, OR SCIENCE

WHAT'S TRAILING BEHIND THAT AIRPLANE?

A long line of clouds often trails behind an aircraft flying at high altitudes. *But are they really clouds or something shadier, more suspicious?* That's the question often asked when it comes to condensation trails, or contrails—not to be confused with or misidentified as fictitious "chemtrails."

When underway, aircraft engines release warm, moist air and tiny particulates from the combustion process. High above, in the cold atmosphere, rapid cooling and condensation occur. Ice crystals quickly form around the particulates and create cirrus clouds. Because the aircraft is in motion, the cirrus clouds extend linearly, creating a contrail. If the atmosphere is dry, it rapidly dissipates and disappears. But if the atmosphere is moist, the contrail may persist and grow long. When the atmosphere is especially moist, contrails may even have fuzzy bottoms. This is precipitation, either rain or ice crystals depending on the temperature. It looks fuzzy because, with dry air below, the precipitation quickly evaporates, thus seeming to vanish as it falls. It's also possible for contrails to form, disappear, and then reappear in a line. This is simply due to the plane flying through pockets of moist and dry air, a common occurrence. The long, linear nature of contrails and their sometimes fuzzy bottoms are particularly distinct from the ground, leading to a heavy dose of suspicion and a wealth of theories regarding "chemtrails."

Chemtrails are a purported means of dispersing chemical or biological agents to modify the weather, cause mass mind control, spread disease, or test vaccinations, as part of a military

SUPERNATURAL, SUSPICIOUS, OR SCIENCE

No one is spraying anything. We promise. It's condensation from engine exhaust.

operation or to reduce world population. But rest assured, the long, linear trails up in the sky associated with aircraft are not dispersal agents. They are simply cirrus clouds forming from engine exhaust. No evidence has been found to prove otherwise, and emissions from aircraft engines are highly regulated to ensure no harm to human populations.

A growing body of research suggests that aircraft exhaust and contrails can contribute to climate change (more in chapter 10). Clouds, including contrails, help to trap heat emitted from Earth, thus adding to warming. This is most prevalent at night because during the day, contrails and other clouds also reflect some of the sun's radiation back to space, providing some cooling. Work is being done to reduce fossil-fuel use in aircraft and to understand how changing flight tracks and timing could reduce the contribution of contrails to climate change.

HOW ABOUT SANDSTORMS?

An enormous cloud of billowing sand rolls over a cliff and sweeps across the land. Within it is the gaping mouth of a mummy that swallows a small airplane—it's the movie magic in 1999's *The Mummy*. Real sandstorms can be terrifying all on their own, even without the help of an out-for-vengeance, returned-from-the-dead creature or some other supernatural force. Once again, it's all about science.

Sandstorms, also known as dust storms or haboobs, can turn day to night, cause deadly pileups on roads, erode buildings, clog machinery, destroy crops, kill livestock, and threaten human health. The ingredients needed to create a particle-laden storm are strong winds, large amounts of sand and/or dust, and bare, dry soils. Overgrazing, drought, wildfires, and vegetation loss increase the likelihood of a sand or dust storm. They can also be kicked off by downdraft winds associated with thunderstorms or a frontal passage.

In general, sand or dust storms occur most commonly in arid and semiarid regions and can suspend and transport particles for hundreds to thousands of miles. Larger particles fall out first, perhaps within hours, but smaller grains can stay aloft for days. Rain can bring relief by washing the dust and sand from the atmosphere. Some dust or sandstorms are so big they're visible from space. Research has shown that massive clouds of Saharan dust blown off Africa can suppress hurricane formation over the North Atlantic Ocean.

Today, choking clouds of sand and dust occur most commonly in the Middle East, North Africa, China, and the southwestern United States. However, thunderstorms with strong downdrafts can pick up dust anywhere conditions allow, but these are usually short-lived events. As the climate warms and areas become increasingly arid, with droughts that are more intense and enduring, extreme dust and sandstorms may become more frequent and occur in new geographic regions.

WHAT THE HECK IS THAT HOVERING BALL OF LIGHT?

The mysterious phenomenon known as ball lightning is what first inspired Dave's lifelong passion for weather. It's not supernatural or suspicious, but it's also not well understood. Turn the page to the next chapter, in which we delve further into the powerful and shocking marvel that is lightning.

■ ■ ■

We provide our sources at the back of the book and invite you to explore and learn more about the fascinating, but not supernatural, topics covered here.

Chapter Seven

LIGHTNING

METEOROLOGIST MIKE SEIDEL might be the king of in-the-field live shots, having done more than a whopping twenty-five thousand. What worries him most during storms? Lightning! Every day across the globe, lightning strikes about 8.6 million times. In the United States each year, some twenty-two million lightning strikes occur, claiming on average twenty-eight lives and causing 243 injuries. Sometimes it is simply bad luck; in other cases, lightning-caused fatalities are the result of poor choices based on misunderstanding and/or a lack of information.

WHAT CAUSES LIGHTNING?

The recipe for lightning includes an unstable air mass, heat, and rising moist air. This is one reason Florida is considered the lightning capital of America. Here's a typical summer scenario in southwestern Florida that could lead to an enormous atmospheric spark or discharge of electricity—lightning.

It's late afternoon on a humid summer day in Tampa. The atmosphere is unstable, and as a result of daytime heating of the land, the overlying moist air has warmed and now rises. As the air climbs higher, it cools and condenses. A cloud grows skyward. More air rises and cools. The cloud billows ever taller and inside, hail or graupel (soft, slightly melted pellets of ice-covered snow) form and swirl about. Within the cloud, the hail or graupel collide with water droplets and ice particles. This creates a charge on the particles, like static electricity when you rub a balloon on your hair. Updrafts ensnare the ice and hail/graupel, with smaller particles rising faster and higher. Positive charges build in the upper region of the cloud, and negative charges collect near the base. As the cloud grows darker and more ominous, the electrical charge builds. Up until now, air has acted like an insulator within the cloud and between the cloud and the negatively charged ground. But soon the charge is too great. Within the cloud, a great electrical discharge occurs, and a bolt of lightning streaks across the sky. Thunder roars. Minutes later, another bolt arcs toward an oppositely charged "step leader" (an advancing upward or downward column of charge) reaching up from a tall palm tree. A crack rings out as lightning strikes the tree, which splits in two and is left smoldering.

Lightning is caused by the buildup of electrical charges in the atmosphere leading to a rapid discharge. A bright white flash is typical, but sometimes moisture in the atmosphere can cause lightning to appear blue or pinkish.

LIGHTNING

DOES IT ALWAYS THUNDER WITH LIGHTNING?

With lightning comes thunder—always. When lightning strikes, it rapidly heats the air temperature to five times hotter than the surface of the sun, that's a scorching 50,000°F. This rapid and extreme heating of the air causes it to instantaneously expand and then contract, generating a sound wave that we hear as thunder. Because light travels faster than sound, thunder is heard after the flash. Thunder can usually be heard up to about ten miles from a lightning strike. If you see lightning but don't hear thunder, it means you are too far away or moisture in the atmosphere has absorbed the sound. This leads to a very common misunderstanding.

HEY, IS THAT HEAT LIGHTNING?

It's a hot summer night and in the distance, lightning streaks across the sky or flashes on the horizon. But there's no thunder. It's heat lightning, right? Nope, no such thing. It's regular old lightning from a thunderstorm. Either you are too far away to hear the thunder or it's been absorbed by moisture in the atmosphere. The ingredients for lightning (an unstable atmosphere, heat, and rising moist air) occur most often in the summer, which probably led to the idea of "heat lightning." But heat alone does not cause thunderstorms or lightning.

CAN WE PREDICT LIGHTNING STRIKES?

Exactly where and when lightning will strike cannot be predicted. However, we can forecast when and where storms are likely to produce lightning, and we can monitor and track lightning as thunderstorms develop and move. In the future, we might be able to do better.

With the Geostationary Lightning Mapper (GLM) sensor aboard the Geostationary Operational Environmental Satellites (GOES-16, GOES-17, GOES-18, and future GEO-XO missions) and some automated intelligence, scientists can now predict the *probability* of lightning striking in areas where thunderstorms are developing or moving. This innovation may one day lead to more precise lightning alerts for apps on mobile devices and for particularly vulnerable outdoor venues, such as a ball field, a golf course, a pool, and the beach. There are also lightning detection networks operating across the United States and world that can alert subscribers to the threat of approaching lightning.

ARE TREES A SAFE PLACE TO TAKE SHELTER IN A THUNDERSTORM?

No! *No!* And *NO!* When lightning strikes, it seeks out the closest connection such as step leaders from tall buildings *and trees*. Trees and other tall objects don't attract lightning; they are just more likely to get struck. When lightning hits a tree, the charge can jump to objects (such as a person) and spread out along the ground, making the area under and around a tree an extremely dangerous place to be.

If you're caught in the open when a thunderstorm approaches, try to find a location where you aren't the tallest object around, but never lie flat on the ground. See the references at the back

LIGHTNING

of the book for outdoor lightning safety guidance from the National Weather Service.

CAN I USE ELECTRICAL APPLIANCES, LIKE A BLOW DRYER, WHEN A THUNDERSTORM IS NEARBY?

No! And don't get in the shower either. Water and electrical lines can act as conduits for lightning if it strikes near your home.

REGULARLY CHECK RADAR FOR STORMS DUE TO LIGHTNING RISK

Ellen could find using her hair dryer during a thunderstorm a shocking experience.

LIGHTNING

HOW ABOUT RUBBER-SOLED SHOES? WILL THEY PROTECT ME?

Sorry, that's another *no*! Let's see, a lightning bolt at 50,000°F and 300 million volts (30,000 amps) versus an inch of rubber at the base of your shoes? Not going to help.

CAN LIGHTNING STRIKE WITHOUT A STORM OVERHEAD?

Sometimes lightning appears to come out of the clear blue sky; this is known as a "bolt from the blue." Arcing out of a thunderstorm, this type of lightning can travel miles away from a storm cloud, angle down, and strike the ground (plate 12). Such lightning strikes have been documented more than twenty-five miles from their source.

If thunderstorms are in the area, even if not directly overhead, it's best to go with "when thunder roars, go indoors." Lightning is particularly dangerous in open areas such as a golf course (especially when holding a metal stick), an open mountain face, a pool or beach, or any open field where sports are played. Although pure water does not conduct electricity, water with ions in it (almost all nonpurified, nondistilled water) does. So, pools and the ocean are no place to be if thunder rumbles. It is best to seek shelter before lightning and thunder start and wait thirty minutes after the last rumble before resuming outdoor activities. It can seem like a painfully long wait, but it's better than the alternative.

To estimate how far lightning is from you, after the flash, count the number of seconds to the crack or rumble of thunder.

Divide this number by five. The result is the distance from the lightning in miles. If the thunder arrives within a minute or less, seek immediate shelter in a safe place such as a building or a covered vehicle.

If your hair stands on end, up toward the sky, lightning may be about to strike. Get indoors as soon as possible and forgo the selfie . . . it could be your last flash.

LIGHTNING NEVER STRIKES THE SAME PLACE TWICE. RIGHT?

Wrong. It can, and it does.

WHAT IS ST. ELMO'S FIRE?

It's not fire, lightning, or dangerous, but this glowing blue phenomenon has been captivating people, particularly mariners, for centuries. Named after the patron saint of sailors, St. Erasmus or St. Elmo, it's most often seen during thunderstorms around ship masts or atop tall, pointy buildings and is similar to what creates the glow in neon signs. It can also happen around aircraft flying through a highly charged atmosphere. The culprit here is plasma, or superheated matter in which the atoms in air molecules have been ripped apart into ionized or charged proton clusters and electrons that glow. What can rip apart air molecules and atoms to create plasma? High voltage . . . like the electrical charge in a thunderstorm. The plasma interacts with nitrogen and oxygen in the atmosphere, causing it to give off a discharge that glows blue or purple—that's St. Elmo's fire.

LIGHTNING

SPRITES, BLUE JETS, AND ELVES?

These short-lived glowing wonders sound like special effects from *Lord of the Rings* or *Top Gun*, but they are real phenomena that occur in the upper atmosphere atop thunderstorms. Sprites, flashes of quick reddish-orange light, occur above cloud-to-ground lightning and are shaped like columns, carrots, or our favorite—jellyfish. Blue jets are unconnected to lighting strikes and appear as brief blue streaks above a thunderstorm. ELVES (Emissions of Light and Very low frequency perturbations due to Electromagnetic pulse Sources) are more diffuse glows or rings of light seen high above active thunderstorms. These phenomena, like St. Elmo's fire, are dazzling but not dangerous.

TWO ELECTRIFYING WORLD RECORDS

Advances in technology have rapidly improved our ability to monitor, track, and record lightning. Using the Geostationary Lightning Mappers (GLMs), mentioned earlier, on the NOAA GOES-R series of satellites, scientists observed two electrifying record-breakers. In April 2020, over the southern United States, a cloud-to-cloud lightning flash streaked some 477 miles across the sky. It was the longest lightning flash ever recorded. In the same year, in June, a single lightning flash lasted a world-record seventeen seconds during a thunderstorm over Uruguay and northern Argentina. Extreme lightning bolts that extend over hundreds of miles are now called megaflashes.

WHAT ARE MY CHANCES OF BEING STRUCK BY LIGHTNING?

The chances of being struck by lightning vary depending on factors such as where you live, what you do, and your choices. For instance, if you live in Florida and play golf in the summer, your chances are greater than those of someone residing in Alaska, where thunderstorms are rare. If your job entails a lot of time outdoors in thunderstorm-prone regions, your chances increase. If you spend time hiking on mountains and like to sit at the summit to watch thunderstorms, your chances of getting struck by lightning are high. In general, the chances of having a personal lightning encounter are low, especially if precautions are taken, but still greater than the odds of winning a single giant Mega Millions or Powerball jackpot. The best way to prevent being struck by lightning is to seek shelter when thunderstorms are in the area.

Lightning is also produced in volcanic eruptions, snowstorms associated with intense low-pressure systems, intense wildfires (which they can also trigger), and strengthening hurricanes. Please note: It's lightning, not "lightening." And remember: When thunder roars, go indoors!

■ ■ ■

See the back of the book for our sources and where you can learn more about lightning.

Chapter Eight

HURRICANES

TROPICAL CYCLONES are one of the most destructive forces on Earth. We and our colleagues are asked related questions that range from basic science and technology to wishful thinking. Some inquiries reveal serious misunderstandings that can put people at risk. We hope the following answers and information will help people better understand tropical cyclones, how they are forecast, their impacts, and when individuals or communities are at risk, to make wise choices.

THE OLD INDIAN MOUND THEORY

Florida is home to a wealth of ancient Indian burial mounds. For some people, they are evidence of an important indigenous past and a rich cultural history. For others, burial mounds are a godsend, offering legendary protection from hurricanes. Let's

Indian mounds do not provide hurricane protection. Period.

nip this one in the bud, like mermaids and megalodons. Ancient Indian burial grounds do *not* provide protection from hurricanes. The evidence is in the tracks—hurricane tracks. Since 1851, 452 hurricanes and tropical storms have crisscrossed the state, including areas hosting ancient burial mounds (plate 13). And remember, hurricanes are much larger than the centerline track. Bottom line: Hurricanes do not magically veer away from or avoid Indian burial grounds, in Florida or elsewhere (same goes for tornadoes).

CAN'T WE JUST NUKE 'EM?

Hurricanes, or cyclones and typhoons as they are known in the Pacific and Indian Oceans, can wreak catastrophic damage. Even weak storms can cause disaster. We cannot prevent hurricanes from forming, alter their course, or stem their power. When it comes to hurricanes, we are at the mercy of a sometimes wicked atmosphere. But on the positive side, hurricanes help to balance the distribution of energy in the atmosphere, in the ocean, and across the planet. Looking for a means to combat and weaken or stop storms, people like to offer "interesting" ideas or sometimes seek someone to blame.

A common inquiry about stopping hurricanes is the classic: *Why can't we just nuke 'em?* The problems here are unintended consequences and collateral damage. The impact of a nuclear blast would reach far beyond "the storm" and be . . . bad, very bad. Dr. Rick Spinrad, the undersecretary of Commerce for oceans and atmosphere and administrator of the National Oceanic and Atmospheric Administration (NOAA), has a thick file of questions and hurricane-related ideas people have presented over the years. He is particularly fond of one: *Can we use all the submarines just lying around to mix and cool the ocean beneath a hurricane thereby stealing its fuel—ocean heat?* Dr. Spinrad actually did the calculations to see how many submarines it would take to adequately stir and cool the ocean underlying a hurricane. The number: 990! That's a no-go for that one.

Ideas to weaken hurricanes presented to Ken Graham, the director of NOAA's National Weather Service and former director of the National Hurricane Center, include towing an iceberg into the Gulf of Mexico to cool the ocean or causing an oil spill to prevent heat transfer. Nope, and nope. The conspiracy-loving public also have some ideas about hurricanes. Director Graham noted these usually involve weather manipulation and range from a means to hide unidentified flying objects (UFOs) to job security for meteorologists—both of which we can confirm with great confidence are not true. Meteorologist Doug Hilderbrand of the National Weather Service has been asked to either confirm or deny that the names for hurricanes added to the list each year are chosen based on connections and/or bribery. We can assure the inquirer and readers that no bribes or conspiracies are involved there either. The World Meteorological Organization (WMO) uses a strict procedure to choose the names for hurricanes in the Atlantic and the eastern North Pacific Basin.

WHAT'S NEEDED TO CREATE A HURRICANE?

The recipe for a hurricane can vary slightly, but the basic ingredients remain the same. An atmospheric disturbance (a pulse of energy or area of low pressure/rising air with disorganized storminess) provides the starter. Add a moist atmosphere, low wind shear, and sufficiently warm ocean waters, and the stage is set for a potential tempest. Because wind shear can blow the tops off developing thunderstorms, essentially tearing a storm apart, the less wind shear the better for a hurricane. Hurricanes obtain their fuel from the underlying warm ocean, but water temperatures must be at least 79°F to a depth of about sixty feet to provide the necessary energy.

HURRICANES

With favorable conditions and the right mix of ingredients, heat from the warm ocean causes air to rise, so that moisture evaporated from the sea is carried upward. As the air rises, more is sucked in from below and surface winds begin to converge. Air continues to warm, pick up moisture from the ocean, and go aloft. A bit higher in the atmosphere, condensation creates clouds and releases heat, which fuels the growth of thunderstorms. As more air converges, rises, and condenses, the storm becomes more organized. Atmospheric pressure drops, and winds begin to spiral around the central low pressure. When a storm's maximum sustained winds exceed seventy-four miles per hour, it's a hurricane.

WHERE DOES THE SPIN COME FROM?

Hurricanes (and low-pressure systems) rotate counterclockwise in the Northern Hemisphere and clockwise in the Southern Hemisphere due to the Coriolis force, which results from Earth's shape and rotation. About every twenty-four hours, Earth spins around once. To complete one rotation in the same amount of time, where the planet's diameter is greater (the equator) it must spin faster. The differing speed of rotation at varying latitudes affects large, unattached masses moving over or taking off from Earth's surface—this is Coriolis. It deflects large moving air or water masses (or missiles) to the right in the Northern Hemisphere and to the left in the Southern Hemisphere. Coriolis is negligible at the equator. To be clear, cars do not veer one way or the other due to Coriolis, and toilets do not flush in different directions on either side of the equator (no matter what tourism operators say or demonstrate).

Because of the lack of Coriolis, hurricanes do not form at the equator. But to the north and south, in two bands between about five and thirty degrees latitude, conditions are ripe for hurricane development, especially in the summer and early fall. While hurricanes (cyclones and typhoons) do occur in the Southern Hemisphere, most form in the Northern Hemisphere, where warmer ocean waters, low wind shear, and pressure disturbances coming off the land set the stage for development. Meteorologists refer to an area in the tropical Atlantic as the main development region (MDR), but when the right conditions exist, hurricanes can and do form anywhere, including very close to the coastline.

HOW GOOD IS THE FORECAST?

Whether on a mobile device, website, television, or radio, we now have access to up-to-date forecasts and weather alerts 24/7/365. But when a hurricane forms, people want to know where it's going, when, and how strong the storm will be. And just how confident are the forecasters in the forecast?

Hurricane forecasts are better today than ever before. This is due in part to an increasing number of observations from around the world provided by satellites such as the United States' GOES-R geostationary series and JPSS polar-orbiting satellites and those from our international partners, along with weather balloons, aircraft (including drones), buoys, Argo floats, drifters, saildrones, and ground-based observing platforms. These data are collected, processed, analyzed, and then integrated into a growing suite of complex and improving numerical models. The models generate visualizations, synoptic overviews, and forecasts that are analyzed by seasoned hurricane forecasters and provided to authorities and private-sector organizations, such as the Weather Channel, local TV stations, and many others. But still, how good are the forecasts?

Since 1990, tropical cyclone (tropical depressions, storms, and hurricanes) track forecasts by the National Hurricane Center (NHC) in the Atlantic have improved by 71 percent three days out and 67 percent one day out ("out" meaning hours and days from a specific location). We hear a lot of numbers like this, but what do they really mean? In 1990, the NHC forecast track for a tropical cyclone three days from its location could be off by an average of about 312 nautical miles. By 2021, advances in forecasting reduced that to eighty-nine nautical miles (a 71 percent reduction in error). Twenty-four hours, or one day, out, the average error went from 107 nautical miles to just thirty-five nautical miles. These are significant improvements; they mean the forecasts were good before but are getting even better. Even small improvements can result in millions of dollars in savings in evacuation-related and preparation costs.

Intensity forecasts, for the most part, are now also improving. In the Atlantic since 1990, hurricane intensity forecasts have improved 29 percent three days out and 11 percent one day out, meaning in 2021 average intensity forecasts were off by an average of 13 mph three days out and only 9 mph one day out.

When hurricanes approach the coast, having track and intensity forecasts that are as accurate as possible is critical to providing the information needed to save lives and property. And these days, they're usually right on the money. But some storms are harder to forecast than others. In 2023, Hurricane Otis proved this point when it went from a weak tropical storm to a Category 5 beast with winds at 165 mph. Over just a twenty-four-hour period, winds increased 115 mph. It was the most extreme case of rapid intensification ever observed, and none of the forecast models or hurricane forecasters predicted it. Acapulco was hit with a catastrophically intense hurricane that wrought damages estimated at ten to sixteen billion dollars.

Hurricane Otis showcased the need for additional work to continue improving the accuracy of forecasts, especially when it comes to rapid intensification.

CAN RAPID INTENSIFICATION BE ACCURATELY FORECAST?

Reliably predicting rapid intensification has been called the holy grail of intensity forecasting. Rapid intensification is defined as when a hurricane's maximum surface winds increase by 35 mph or more within twenty-four hours. The processes involved are highly complex, range widely in scale, and are affected by both internal and external factors. Internal factors include the behavior of clusters of intense thunderstorms, the tilt of a storm's vertical structure, changes in atmospheric pressure, and eyewall replacement cycles. External influences include wind shear, upper ocean temperature (ocean heat content), salinity, atmospheric moisture, and interaction with other weather systems or longer-term cycles in the coupled atmosphere-ocean system. In other words, it's very complicated.

One now-notorious cause of rapid intensification is when a storm moves over an area of particularly warm ocean water. In 2023, extremely warm ocean waters near the coast of Mexico are thought to have contributed to the rapid intensification of Hurricane Otis. In the past, the high-test fuel needed for rapid intensification has also been provided by the Loop Current and associated warm core eddies in the Gulf of Mexico or the Gulf Stream. Hurricanes that intensified rapidly while moving over the Gulf of Mexico include Opal (1995), Katrina (2005), Harvey (2017), and Ian (2022); the Gulf Stream fueled Hugo (1989) and Andrew (1992). But—and here is another big but—if the warm water is not deep enough, a passing hurricane may not rapidly intensify.

Much of our information on ocean temperature comes from sensors aboard satellites, but these detect temperature only at the very surface of the sea. Additional data from buoys, gliders, Argo floats, and other ocean-observing platforms are needed to reliably predict rapid intensification—adding to the challenge of forecasting it.

With higher-resolution computer models that couple the ocean and the atmosphere, NOAA's newest satellites, and data from subsurface oceangoing buoys and uncrewed gliders, along with the heroic efforts of the NOAA and U.S. Air Force Hurricane Hunters, our ability to forecast rapid intensification is improving. Ocean-observing platforms provide important input and the verification needed for higher-resolution models. This is critical for storms approaching the coast, like

Hurricane Otis. Rapidly intensifying hurricanes close to shore put populations at great risk and can create serious challenges in preparation and evacuations. Ask a meteorologist or a hurricane forecaster for a nightmare scenario, and you'll get the following: waking up to a hurricane that was a tropical storm at bedtime and in the morning is a Category 4 or higher and about to make landfall.

WATCHES VERSUS WARNINGS: WHAT'S THE DIFFERENCE?

NOAA's National Hurricane Center issues official hurricane watches and warnings. A hurricane watch is issued when hurricane conditions (sustained winds of 74 mph or greater) are *possible* somewhere within a specified area. A hurricane warning is issued when hurricane conditions are *expected* somewhere within the specified area. Preparations should begin if a hurricane watch is issued for your area. With a hurricane warning, you should complete storm preparations and evacuate if needed. And please, if directed by local authorities to evacuate, don't wait. Go as soon as possible.

THE CONE OF UNCERTAINTY AND CONFUSION

For many people, the track forecast cone, or "cone of uncertainty," could be called the cone of confusion. Following Hurricane Ian in 2022, scientists specifically investigated what people found most confusing about the forecast cone. Results showed that the size of the storm and where damage was likely to occur were the most misunderstood. Because . . . the cone of uncertainty *does not include* information about a storm's size (wind field) or potential impacts, such as expected precipitation, storm surge, or flooding. Using the cone to interpret storm size and/or potential impacts leads to confusion as well as poor and unsafe decision making. The deputy director of the NHC even delivered an impassioned talk at the 2023 National Weather Association's annual conference to drop using the cone of uncertainty and focus more on impacts.

In addition to the forecast cone, the National Hurricane Center does provide the forecast wind field, wind-speed probabilities, and when tropical storm–force winds are most likely to arrive, along with watches and warnings, expected peak storm surge, and probable storm-surge inundation based on the latest forecast (plate 14). In association with the Weather Prediction Center, predicted inland flooding is also provided, as is the expected damage from strong winds.

Another point of confusion about the forecast cone comes from focusing on the centerline track and a storm's predicted landfall, which together appear as a line and a point. But when a tropical cyclone makes landfall, the impacts extend beyond, many times far beyond, where the center comes ashore. Impacts can be extreme from the center out to hundreds of miles or more—it's not a point of land. And if an unexpected wobble or jog in a storm's track occurs (and they will), the landfall or area of impact can change quickly, especially if a storm is moving parallel to shore. *So, what does the cone of uncertainty really mean?*

The National Hurricane Center (NHC) track-forecast cone for developing tropical cyclones extends from the current

PLATE 1 While mermaids are mythical, imaginary, and fun, this photo is not scientific evidence that they exist.

Photo from Atlantis, Annapolis, MD (not the other Atlantis) by Dave Jones

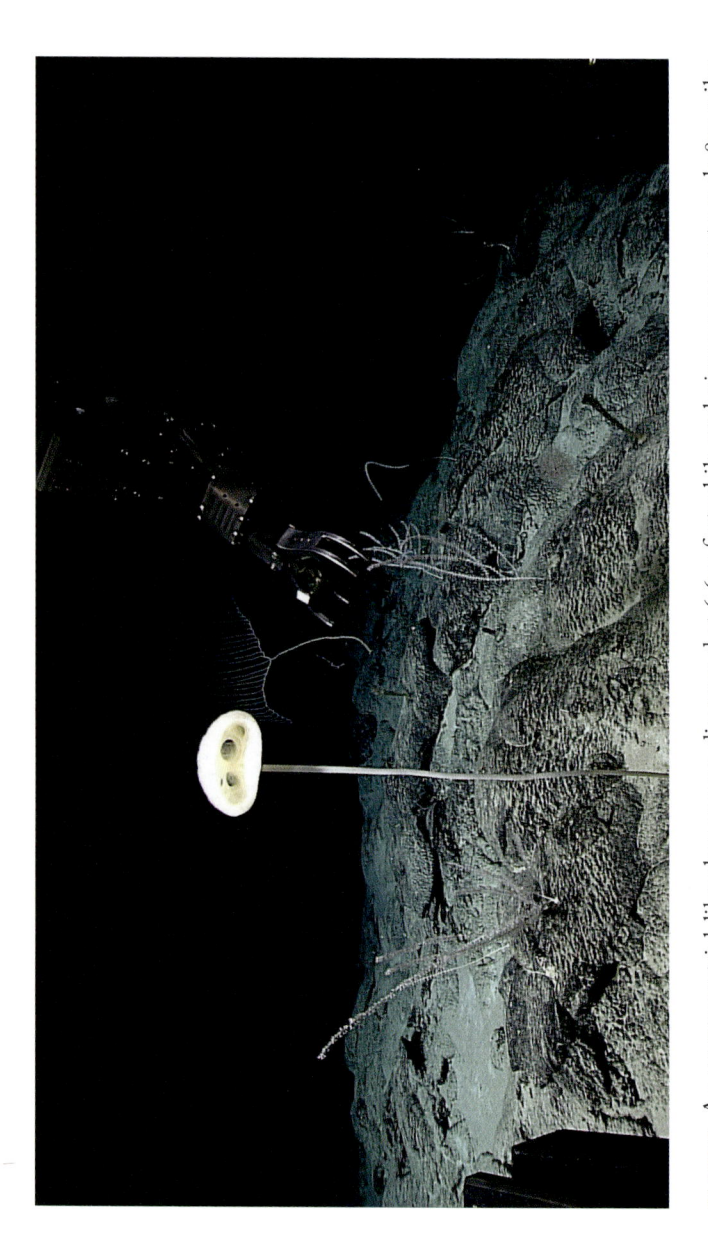

PLATE 2 An extraterrestrial-like glass sponge discovered at 6,650 feet while exploring a seamount nearly 850 miles southwest of Hawaii.

Courtesy of NOAA Office of Ocean Exploration and Research

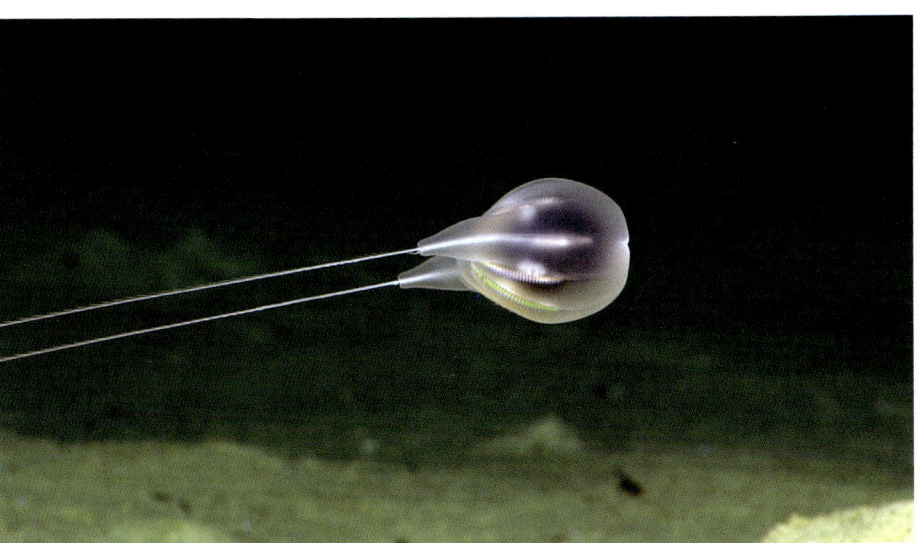

PLATE 3 (*a*) Casper, the unusual ghostlike octopus discovered in the Hawaiian archipelago. (*b*) A deep-sea, hot-air-balloon-like comb jelly found off Puerto Rico.

Photos courtesy of NOAA Office of Ocean Exploration and Research

PLATE 4 A hammerhead shark. Check out its countershading coloration, dark on top and light underneath. Its bar-shaped head provides an undersea rearview mirror and expanded electrosensing capabilities.
Courtesy of Stephen Frink

PLATE 5 Joodles (jellyfish noodles) as eaten by us in (*a*) New York and (*b*) a Florida restaurant. We attest that, when prepared well, they are a tasty and surprisingly nonslimy treat.
Photos by Ellen Prager

PLATE 6 These sea turtles aren't just hanging around; they're harmlessly harnessed up to study their orientation response to regional magnetic fields. (*a*) A baby (hatchling) loggerhead sea turtle. (*b*) A larger, juvenile green sea turtle.

Courtesy of Dr. Ken Lohmann/University of North Carolina.

PLATE 7 An unprecedented heat wave hit the ocean in 2023. Corals throughout the Florida Keys bleached and died. (*a*) Bleached elkhorn coral on North Dry Rocks Reef, off Key Largo, in September 2023; (*b*) close-up of bleached coral polyps on Alina's Reef, off Key Largo, also in September 2023.

(*a*) Courtesy of Liv Williamson; (*b*) courtesy of Alexandra Wen/University of Miami

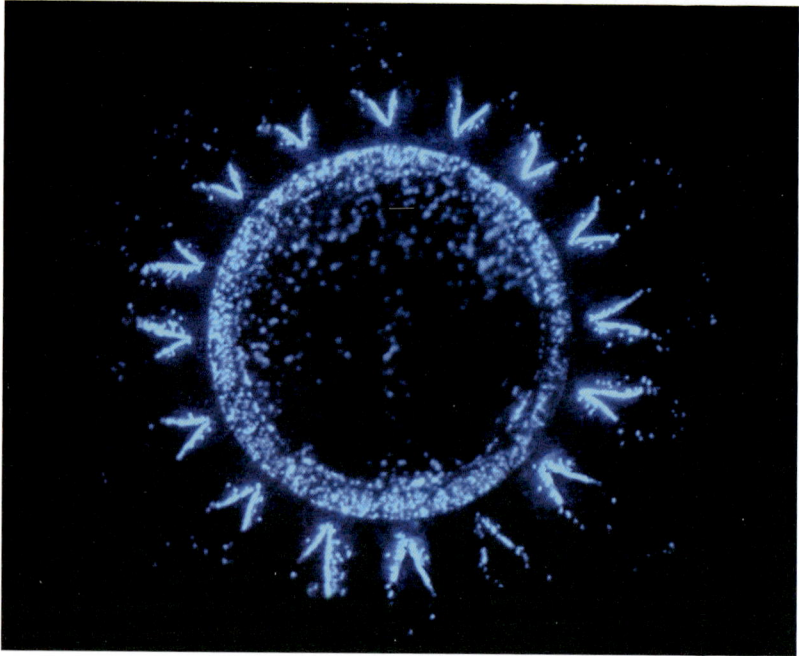

PLATE 8 Bioluminescence is the ocean's own glow-in-the-dark spectacle. (*a*) The dinoflagellate *Pyrodinium bahamense*, disturbed when a ladder splashes down from a dock, lights up the night in Indian River Lagoon, Florida. (*b*) A deep-sea crown jellyfish bioluminesces in the lab after collection using the Johnson Sea Link submersible.

Courtesy of Dr. Edie Widder/Ocean Research and Conservation Association

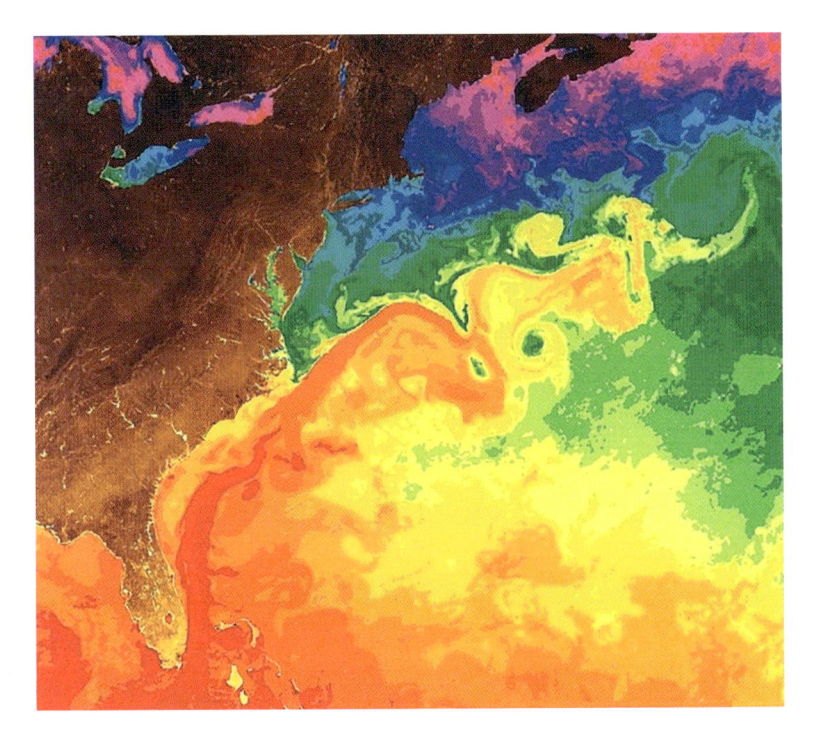

PLATE 9 Sea surface temperature showing the warm Gulf Stream (*red*) flowing north off the southeastern U.S. coast. Note the cold-core eddies, or rings, to the south of the Gulf Stream and warm-core eddies to the north.
Courtesy of NASA

PLATE 10 Up in the sky! It's hovering, it's saucer shaped, it's a lenticular cloud! Photographed over Fort Collins, Colorado, in 2020. Colorado is one of Dave's favorite places to see cool clouds.
Photo by Walter Lyons

PLATE 11 Cowabunga! Sorry, no surfing here. These are wave or Kelvin-Helmholz clouds over Fort Collins, Colorado, in 2019.
Photo by Walter Lyons

PLATE 12 Lightning can strike miles away from a thunderstorm. It's known as a "bolt from the blue." When thunder roars, go indoors! This storm was in 2016 near Cabo Rojo, Puerto Rico.
Photo by Frankie Lucena

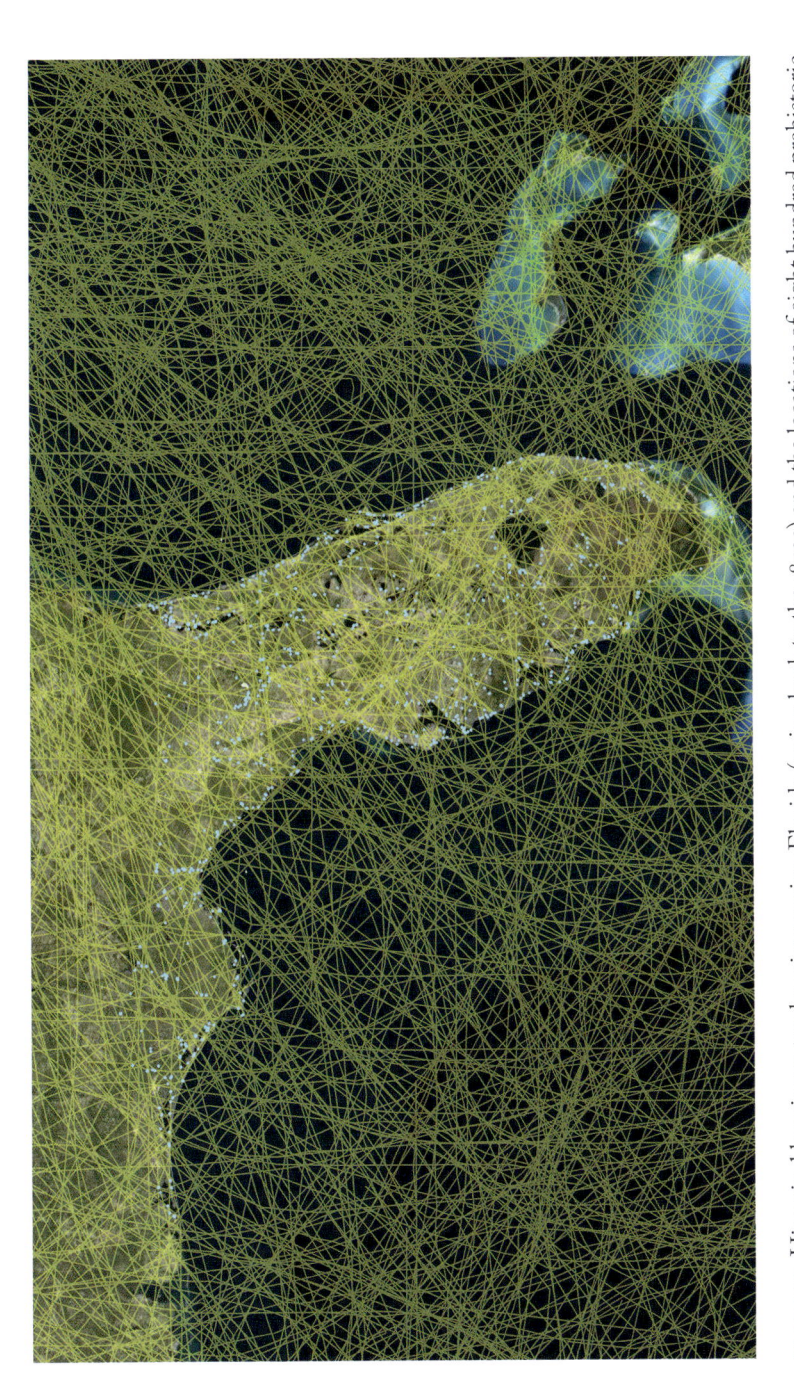

PLATE 13 Historical hurricane tracks crisscrossing Florida (going back to the 1840s) and the locations of eight hundred prehistoric burial mounds (blue dots). Approximately 484 sites are within less than one mile of a storm track. And remember, hurricanes can be hundreds of miles wide!

Courtesy of Dr. Thomas Pluckhahn/University of South Florida

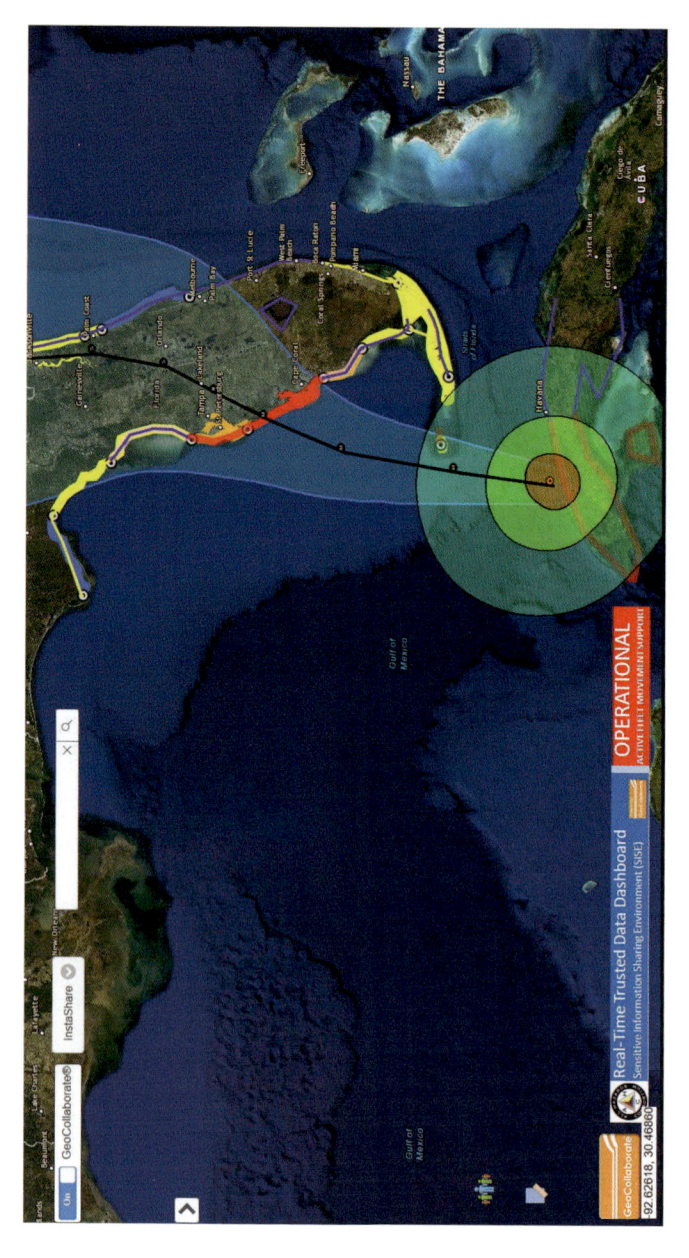

PLATE 14 The National Hurricane Center (NHC) forecast cone for Hurricane Ian on September 27, 2022, at 11 a.m. EDT shown in GeoCollaborate. Additional information shown includes tropical storm warnings (*purple*), hurricane warnings (*red*), peak storm surge forecast, and the size of Ian's wind field.

Courtesy of Dave Jones

PLATE 15 Why broadcast meteorologists can't show every town's temperature on a local weather map. If they did, it would look like this.
Courtesy of Dave Jones

PLATE 16 Rising sea levels due to climate change cause more frequent flooding and higher storm surges, leading to big-time coastal threats, decreasing coastal real estate values, and skyrocketing insurance rates. Someone we know photoshopped a little dose of reality onto this "For Sale" sign in St. Petersburg, Florida.
Photo by Ellen Prager

PLATE 17 (*a*) Sunrise in Norfolk, Virginia, where rising sea level is spurring big plans to combat more frequent flooding in town and at Naval Station Norfolk, the largest naval complex in the world. (*b*) Sunny-day flooding during a king tide, October 30, 2023. Now a school, the Hague was once the Unitarian Church of Norfolk until they evacuated.

Photos by Dave Jones

PLATE 18 A sun dog in Leesburg, Virginia, 2023. Ice crystals in cirrus clouds to the left of the sun refract the light, producing a beautiful rainbow effect.

Photo by Dave Jones

PLATE 19 The Sun's luminous atmosphere, or corona, as seen during the total solar eclipse on August 21, 2017, above Madras, Oregon. This is the only time during an eclipse that it is safe to take off your protective glasses.

Courtesy of Aubrey Gemignani/NASA

PLATE 20 Holy aurora from space! Photo taken from the International Space Station by astronaut Scott Kelly over the Pacific Northwest on January 20, 2016. This is an example of space weather caused by solar particles interacting with Earth's magnetic field.
Courtesy of ESA/NASA

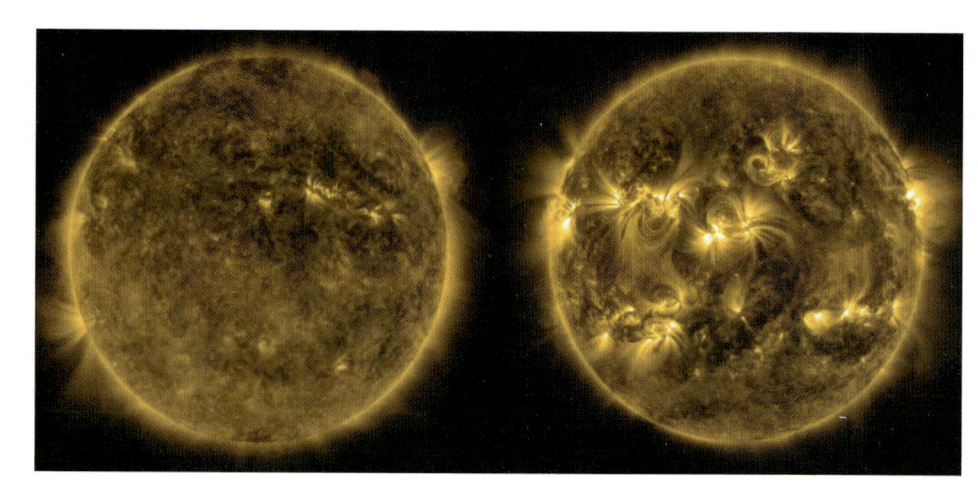

PLATE 21 Side-by-side still images from videos of the Sun during (*a*) low activity, October 2010, and (*b*) high activity, May 2013.
Courtesy of NASA/Solar Dynamics Observatory

position of the storm to the probable location of its center for the following 12, 24, 36, 48, 72, 96, and 120 hours (five days), with an error of about 30 percent (one-third). In other words, based on the data available, model outputs, and historical track records, the center of the storm has a 60 to 70 percent chance of being within the cone. It also means the center of the storm could be outside the forecast track about 30 percent of the time! Where the cone is wider, there is more uncertainty in the forecast track.

Small dots provide the storm's forecast location at each point in the cone. A letter inside the dot indicates whether the storm is, or is forecasted to be, a tropical depression (D), a tropical storm (S), a hurricane (H, Cat 1 or 2), or a major Cat 3+ hurricane (M). If you are anywhere in the cone of uncertainly, you have a 60 to 70 percent chance that the center of the storm will make a direct hit. But again, the cone alone does not provide information on the size of the wind field, the storm surge, the amount of precipitation, and most importantly, the potential impacts. Look to the NHC for additional, critical information to learn about and prepare for a tropical cyclone and to your local NWS forecast office for detailed impact-based information.

WHAT ABOUT ALL THOSE MODELS? WHICH IS BEST FOR ME?

Experts at the National Weather Service and its partners use multiple global and hurricane-centered models to provide guidance and input into their forecasting process. They're run and rerun with varying sets of initial parameters and changing atmospheric and ocean conditions. The result is a set of "spaghetti" plots showing the forecast tracks of low-pressure systems and tropical cyclones based on each run of a model or ensemble of models. Ensembles are models grouped together to generate a higher-confidence forecast. NWS director Ken Graham relayed his frustration with people who shop for the model that agrees with what they want to happen or go with the one they saw first, ignoring updates. With changing conditions and geography, the model or models that provide the best forecast can (and do) change. This is one reason numerous models and ensembles are run repeatedly—and why it is so important to monitor updates as storms develop, move, and approach shore.

Today, information is available from a multitude of easily accessible sources, some reliable and others . . . not so much. From social media to YouTube posts or local reporters providing their own interpretation of the models, the varying array of storm forecasts can be messy and confusing. Just because someone has millions of followers on TikTok doesn't mean they have the expertise to accurately interpret model results and provide a credible hurricane forecast. As "click-bait," unscrupulous posters will upload a model forecast out in fantasyland (like 384 hours out) to rile up followers. It's a disservice to all and unnecessarily generates fear, hype, and misinformation. We suggest identifying a few credible, trustworthy sources and sticking with them. Don't fall for the video that touts a Cat 6 will hit Somewheresville with an end-of-the-world storm surge. That's a waste of your time, energy, and mental well-being—and soooo frustrating to those who provide trustworthy updates.

The National Hurricane Center official forecast remains the most reliable information available, with NWS forecast offices providing higher detail for local areas. Your favorite local broadcast meteorologist can also be a trusted source for storm information, as can someone else you have researched who provides credible weather and climate information.

ARE CAT 1 AND CAT 2 HURRICANES SAFE TO RIDE OUT?

In the 1970s, wind engineer Herb Saffir and meteorologist Bob Simpson wanted to help alert the public about potential hurricane damage. They created what today is the Saffir-Simpson Hurricane Wind Scale, describing the intensity of a hurricane, ranging from a dangerous Category 1 to a catastrophic Category 5, based on the maximum sustained wind speed at the surface (peak one-minute wind at a height of thirty-two feet). This scale, however, does not account for the size of the wind field, the storm surge, the amount of precipitation possible, the angle of approach, the speed of movement, or preexisting or related conditions, such as saturated soils. In other words, the category alone does not provide all the critical information needed to make wise choices before and during a storm.

We often hear people say that if a Cat 1 or 2 is expected, they will ride out the storm. This could be an extremely bad if not deadly decision. Lower-category hurricanes with a wide wind field can result in a significant and dangerous storm surge. Strong winds blowing onshore or alongshore winds over a long distance,

HURRICANES

or fetch, can cause water to pile up on the coast, flow into bays and inland waterways, flood roadways, and rush into homes!

When hurricane Sandy struck New Jersey in 2012, it wasn't even (officially) a Cat 1 storm, but it was enormous. Tropical storm–force winds extended more than 460 miles out from the center. That's almost a thousand miles across! As the storm moved north, parallel to the U.S. east coast, water levels rose from Florida to Maine. To make things worse, the storm hit at a full-moon high tide. On Kings Point, Long Island, the sea rose more than eleven feet above normal. Some beaches lost forty feet of shore, fires ignited from ruptured gas mains, and seawater surged up New York Bay, backing up in the Hudson River and flooding subway tunnels. Sandy's toll in the United States included seventy-two fatalities (forty-one due to storm surge), 650,000 homes destroyed, and thousands of people left homeless.

Hurricane Harvey in 2017 struck the coast of Texas as a Cat 4 storm, but the worst of the impacts came after it was downgraded to a tropical storm. As with Allison sixteen years earlier, the problem was not the wind or the surge but a lingering storm and massive amounts of rain. For days, bands of rain born from unusually warm waters in the Gulf of Mexico drenched Harris County and Houston. So much rain fell that the flooded ground acted like the ocean, providing fuel for the storm. An estimated one trillion gallons of water deluged Harris County. Flooding in and around Houston was exacerbated by dense development on a floodplain and extensive pavement, which funneled water into rivers and low-lying areas.

HURRICANES

Understanding all the potential impacts of an approaching hurricane is vital to making wise decisions that can save lives and property. It's not just about the category; even Cat 1 and Cat 2 hurricanes can be dangerous and deadly. There's also the chance of rapid intensification. And one more thing: There is no such thing as a Cat 6 hurricane (for now). While some people argue there should be a Cat 6, we're not sure it would be all that helpful. If Cat 5 is already designated as catastrophic, what would Cat 6 be? An apocalypse? Armageddon?

HOW DANGEROUS IS STORM SURGE?

In 2022, when Hurricane Ian headed toward the southwest Florida coast (and even afterward), we were asked by people in its path if going into a small interior room would be a safe place to shelter in their single-level home. When strong winds are forecast, and evacuation is not possible, this is often the recommendation. But if the danger is storm surge, taking shelter in an interior room is potentially a fatal mistake, especially if the home has only one level. It is best to identify a designated shelter ahead of time and go there when officials advise.

Storm surge occurs when wind and waves push water ashore faster than it can drain away. The underlying bulge of sea level associated with an approaching atmospheric low pressure, high tide, and waves can add to the water rise. Other factors that influence storm surge include a hurricane's intensity, speed of movement (slow-moving and lingering storms can generate higher surge), the direction of approach, the size of the wind field, the nature of the coastline, sea level, and coastal and offshore bathymetry.

Sophisticated computer models can now predict probable storm surge, but people often ask: *How high will the water come relative to where I live or where I park my car?* Previously, storm-surge forecasts were made relative to mean low water, high tide, or sea level. But few people know where those levels are relative to where they live or work. To make storm-surge warnings more meaningful to more people, NHC forecasters now provide potential surge levels relative to "normally dry ground." Even this description can be confusing. Maps and animations are increasingly available to show inundation levels due to flooding, sea-level rise, and storm surge. These can help people better understand their level of risk, because when storm surge comes to call, it should be taken very seriously.

Moving water is incredibly powerful. A foot and a half of moving water can sweep a person off their feet. In a couple of feet, cars start bobbing. In a hurricane, storm surge sweeps away cars, destroys homes, and can toss boats atop buildings, other boats, or trees. Storm surge is responsible for most hurricane-related deaths. In Hurricane Katrina (2005), a twenty-six-foot storm surge rushed miles inland, ravaging the Mississippi coast. In Louisiana, storm surge broke levees and flooded much of New Orleans. Of the 1,800 fatalities in Katrina, most were due to storm surge. In Hurricane Ian, forty-one lives were taken by storm surge. If significant storm surge is predicted and you are in its path—get out. Do not plan to go into an interior room as your safe shelter—evacuate. And don't wait until it's too late or too dangerous to go. Today, because of climate change, rapid intensification of storms is becoming more common, and when it happens close to shore, the dangers can escalate quickly.

WHAT IMPACT IS CLIMATE CHANGE HAVING ON HURRICANES?

Climate change does not cause hurricanes, and data do not indicate there will be more hurricanes as temperatures rise. Empirical evidence and research, however, show that climate change is causing storms to be stronger and to rapidly intensify more frequently. With higher ocean temperatures, there's more fuel (ocean heat content) to power stronger hurricanes. More moisture in the atmosphere means heavier, more intense rainfalls. And with sea-level rise, storm surge will be higher. As ocean waters warm, hurricane-friendly conditions will also extend increasingly northward, bringing strong storms to new or less frequently impacted regions. It's unclear if and how tracks are changing, if atmospheric steering will cause storms to move more slowly and linger more often, and if in the future the hurricane season will be longer. Bottom line: There may not be more hurricanes in a warming world, but the ones that do form are more likely to be stronger and more destructive, and they may impact more northerly areas. And it takes only one for catastrophe to strike. For more on climate change, see chapter 10.

SHOULD I EVACUATE?

Several of our friends and colleagues have stayed in their homes during major hurricanes, such as Hugo and Andrew. They all agree: Never again. One colleague's house blew apart as he and his family crawled out on their knees to a neighbor's home.

Another person says the noise and shaking were etched into his very being—something he never, ever wants to relive. The decision to evacuate is not simple or easy, but it can be lifesaving.

Critical to making the decision to evacuate is access to reliable and timely information, such as a storm's expected impacts, which are the result of numerous factors, including the size of the storm, its speed of movement, its forecast track and intensity. Understanding the expected impacts and your risk is critical to making wise decisions. It is also important to find out, in advance, what evacuation zone you live in and be prepared if an evacuation is ordered (voluntary or mandatory). But even with all the right information about the hazards involved, a person must determine their own vulnerability and what level of risk is acceptable. For people who have not been through a hurricane, it is often difficult to comprehend the real dangers involved.

If a hurricane is approaching and there's a 90 percent chance of your home being flooded, would you leave? Probably. What if the risk was only 20 percent—would you still go? Twenty percent doesn't seem like a lot of risk, but when the risk is life-threatening, is any risk acceptable? Imagine going to a restaurant. You learn that the health inspector has recently visited and now says there is a 20 percent chance of getting food poisoning at that restaurant. Would you still go? Probably not. But when leaving a home, it's different. There's a loss of comfort and a feeling of safety in your own house. There's fear of not being there if something happens. And there could be logistical, physical, or financial issues involved. But if the forecast warrants it or authorities order an evacuation, it's best to leave and leave early. Do not play chicken with a hurricane—ever.

Waiting for the last minute to evacuate can lead to getting stuck on the highway in the runup to or during a storm—and that is not a good place to be. If a storm is approaching parallel to the coast, a short jog or wobble can swiftly change where the worst impacts occur, leaving no time for evacuation. And don't forget to gas up well before thousands of others have the same idea.

Remember to bring important documents, medications, and what you will need for at least a week, if not more. Hopefully, your home will be spared, and you can return quickly. But with extreme impacts, it could be weeks until you are able to return safely to assess the damage. Above all else, be safe and be smart by basing your decisions on trusted information sources. When it comes to evacuations, you and your family are more important than a structure or material things.

■ ■ ■

Our sources are provided at the back of the book, and we encourage you to learn more about hurricanes.

Chapter Nine

WEATHER FORECASTING AND EXTREME EVENTS

FROM YOUR LOCAL TELEVISION METEOROLOGIST to storm experts on national weather networks, people have a love-hate relationship with their weather forecasters. When the forecast helps us prepare for a storm or successfully predicts a perfect day for a special occasion, forecasters are the best. But when bad weather strikes, or is not what was called for, your favorite or not-so-favorite-anymore meteorologist is to blame. But maybe the problem is not the forecast or the meteorologist. Maybe it's confusion about what the forecast really means, when it was heard, or where the information comes from. And don't expect to learn about meteors in the daily forecast. Meteorologists study the weather; look to astronomers and astrophysicists for more about meteors, comets, and asteroids.

The accuracy of daily weather forecasts has actually greatly improved.

WHY IS MY LOCAL FORECAST FOR RAIN SO OFTEN WRONG?

Your local meteorologist forecasts an 80 percent chance of rain in the afternoon. Where you live there wasn't a drop. Were they wrong? Maybe . . . but maybe not. Forecasts are based on the statistical probably of 0.01 inch or more of precipitation at any point within the forecast area in a specified time interval. In other words, the 80 percent chance of rain forecast meant that during the afternoon (noon to 6 P.M.) there was an 80 percent chance for rain *at any spot* in the forecast area. It doesn't mean that 80 percent of the forecast area will get rain or that it will rain 80 percent of the time in the forecast area. Most likely it did rain, just not in your location. So the forecast may have been correct even though it seemed wrong.

IS IT PARTLY SUNNY OR PARTLY CLOUDY?

If you want to get technical, both are official descriptions for specific conditions. During the day, when between three- and five-eighths of the sky is covered with opaque clouds, it is officially partly sunny. At night, with the same cloud cover, it is partly cloudy. Now you know.

MY WEATHER APP IS GOOD ENOUGH. RIGHT?

Thousands of weather apps are now available on the internet and for mobile devices—some good, some not so good. It's super handy to look at your weather app whenever and wherever you want, but convenience doesn't always equal accuracy. If your weather app says that in forty-eight hours, at 4 P.M., for your pinpoint location, the temperature will be 72.8°F—not so good. Many weather app forecasts are deterministic, meaning they provide a definitive forecast, which makes them seem super accurate. It's also super impossible. Most meteorologists provide forecasts that are probabilistic, supplying information on chances one way or the other. They don't give a definitive down-to-the-decimal or minute-by-minute forecast. Credible weather forecasts are not that precise. So please, don't harass or blame your local meteorologist when an impossibly exact weather app forecast is wrong. Take those with a large chunk of salt, or find another source of information.

While weather forecasting is now more accurate than ever before, it still has limitations. According to NOAA, the National Weather Service official forecast can accurately predict the weather out seven days (the week ahead) about 80 percent of the time. For five days ahead, the forecast is even better, with about 90 percent accuracy. But extend the outlook to ten days or longer and the chances of an accurate forecast are 50–50. That's the same as a flip of the coin.

Will we ever be able to reliably provide that on-time to-one-decimal highly precise temperature forecast ten days out? Probably not. The weather is driven by Earth's constantly changing atmosphere. To forecast it for the days ahead, meteorologists must use numerical weather-prediction models based on our understanding of the atmosphere along with inputs on its current state (initial conditions) from satellite data, surface observations, radar, radiosondes, buoys, drifters, ocean gliders, Argo floats, aircraft, wind profilers, and more. Models must be repeatedly run with changing inputs from validated observations. The more quality data available as input, the better the output. But observations will never be always available everywhere, and even a small perturbation at one location or slight changes in conditions such as wind, pressure, or temperature can change the output.

Weather models are also based on grid points, which dictate the resolution of the output. If the boxes—the spaces between grid points—are small, resolution is high, and vice versa. Running small-grid-size models requires a ton of computer power. Such high-resolution models are best used for local forecasts over a relatively short, one- to two-day, period. But most weather models are regional or global. They have a larger grid size and provide reliable forecasts with lower spatial resolution for longer time periods (a week). This is where the local meteorologist comes in.

Based on weather model outputs, meteorologists at the NWS Weather Forecast Offices (WFOs) issue an official forecast for their county warning area. A county warning area is essentially their area of responsibility. There are 122 NWS WFOs across the United States. Nongovernmental meteorologists look at a myriad of computer forecast models from public and private sources, examine the NWS forecast, and make adjustments based on local and regional factors, such as being in a coastal zone or near a river, mountain, or lake. They will also add specific impacts that are relevant to their customers beyond what the local NWS forecast provides. It's important to note that weather app providers don't generate the official forecast. Some may pass on the NWS forecast in a different format, while others base their forecasts on their own proprietary deterministic weather models.

With the atmosphere in constant flux at all levels, precise by-the-minute forecasts for the days and weeks ahead are not possible and may never be. So, be skeptical if your weather app predicts that tomorrow's temperature will be 56.9°F at 1 p.m. and rain will fall at exactly 4:12 P.M.

WHY DO WE EVEN NEED THE NATIONAL WEATHER SERVICE?

With a wealth of weather apps and private providers and the thinking that it would save the nation gobs of money to do away with the NWS and NOAA, we've heard this question all too frequently. But it's kinda like asking why do we need cows when there's plenty of milk at the store? Or why do we need farms when I can order a salad at a restaurant? Accurate weather forecasts are based on modeling and many, many, many observations including from billion-dollar satellites and international partnerships. Official forecasts use the best available numerical models, which are used by and in many cases run by the NWS. Essential to these models, and others that may be run by private weather providers, are an abundance of observations for input, calibration, and validation. Guess where most of those comes from? Yes, NOAA and the National Weather Service. Don't forget about official watches, warnings, and alerts—the National Weather Service. Next time you hear this question, please let the inquirer know that the NWS and NOAA are critical to our well-being, productivity, safety, and at the cost of a burger and fries per person per year. To consider doing away with the NWS and NOAA is a fast trip down the lane to crazytown and catastrophe.

WHAT'S THE OFFICIAL FORECAST, WATCH, OR WARNING?

NOAA's National Weather Service (NWS) issues official forecasts from the U.S. government; other forecasters, such as your local broadcast meteorologist, may not issue the same forecast. However, only the National Weather Service can issue official watches (hazardous weather is possible) and warnings (hazardous weather is occurring, imminent, or likely and poses a threat to people and/or property).

WEATHER FORECASTING AND EXTREME EVENTS

WHAT WEATHER DO YOU HAVE SCHEDULED FOR MY DAUGHTER'S WEDDING SIX MONTHS FROM NOW?

That was a real question from a viewer when Dave worked as a broadcast meteorologist at NBC4 in Washington, DC. Apparently, this person was such a dedicated follower of NBC4, they thought the meteorologists at the station were never wrong, so they must schedule the weather. If only we could. But alas, sorry, that's a no, weather forecasting doesn't work like that. As

Accurate long-term weather forecasts exist only in the land of make-believe.

for a credible detailed forecast six months out or even thirty days ahead—that's like taking another trip to fantasyland.

THE FARMER'S ALMANAC

While we're on the topic of reliability and long-range weather forecasts, how about the forecasts in the *Farmers' Almanac* and the *Old Farmer's Almanac*? Meteorologist Dan Satterfield put it this way in a 2013 American Geophysical Union (AGU) blog about a winter forecast: "To say it's patently ridiculous would be a severe understatement, because there is NO evidence that 'based on planetary positions, sunspots and lunar cycles' one can forecast the winter weather in North America. NONE, Nada, Zip, and ZILCH. Have I made my point yet??" What's his beef with their forecasts? For one, they are based on a "secret formula," devised in 1792 by the almanac's founder, using many factors including the moon and sunspots. Note the "secret" part. There's no transparency, no peer review. Who knows what's in the magic formula? It could be science, pseudoscience, astrology, or pure we-made-it-up stuff. *Popular Mechanics* author Bishop Rollins also has issues with their stated reliability: "The almanacs say they can predict weather with around 80 percent accuracy, but a University of Illinois study disagreed, saying the Almanac was only about 52 percent accurate, which is essentially random chance."

Across the United States (and world), the summer of 2023 brought record breaking, dangerous heat for days and weeks on end, along with extreme precipitation events that caused catastrophic flooding. We decided to look at the almanac's forecast and see how it did. The headline looked good: "Summer Forecast 2023: Sizzles Return." Then again, in the context of climate change and a developing El Niño, would anyone have gone with "Summer Forecast 2023: The Chill Is On"?

How about for individual states or regions in the United States: how did the almanac do there? As forecast for the summer in Texas, it was indeed sweltering and stormy. Is it ever not sweltering and stormy in the summer in Texas? What about the record-breaking and disastrous storms in the Northeast? The almanac forecast called for scorching and dry. Dry? That's a big oops. Forecasting conditions that generally occur can make the almanac's forecasts appear spot-on, but there's an equal chance they'll be dead wrong. That's a 50–50 in reliability—no better than asking a groundhog to see its shadow, asking our dog to choose from a basket of bones representing weather conditions, or playing darts on a weather-forecasting board. Okay, maybe better than that last one—one of us sucks at darts.

Don't get us wrong. Almanacs provide good information on things such as moon phase, tides, gardening, and more. But even with sophisticated science-based computer models using inputs from observations around the world, the weather cannot be accurately predicted more than about a week ahead. There is no secret formula that can do better. For now, if you want to know what the weather will be more than a week out, there's still an even chance the best forecast will be wrong (or right). NOAA's Climate Prediction Center (CPC) in College Park, Maryland, does issue seasonal outlooks, but don't look for an exact forecast, just the probability or chances of above- or below-average temperatures and precipitation.

CAN FOG PREDICT RAIN?

Tanja Fransen, NOAA's NWS meteorologist-in-charge in Portland, Oregon, tells us that when she worked in Montana, people often suggested that dense fog is a good predictor of significant precipitation ahead—specifically, that ninety days after a dense fog event there will be a large rain event. The time frame given varies: Some people say rain will come in sixty days; others suggest it will be in six or nine weeks. But is fog really a predictor, or is the connection something else—like maybe the season? In the region, fog typically occurs in midwinter. Ninety days after midwinter is . . . spring. And what is the wettest time of the year? Spring! With or without fog, the rains come.

WHY ISN'T MY TOWN'S TEMPERATURE SHOWN ON THE TV MAP?

Every day, several times a day, broadcast meteorologists provide weather forecasts and highlight temperatures in towns across the nation. Even in small localities, if every town's temperature were shown, the maps would be totally unreadable (plate 15). This is why only a limited number of temperature locations are highlighted each day on the forecast map. Most broadcast meteorologists highlight areas when data become available, and they like to show towns where "weather watchers" provide observations on their local conditions. To improve the chances of getting an on-air mention for your town, become a weather watcher.

WEATHER FORECASTING AND EXTREME EVENTS

WHAT THE HAIL? OR IS IT SLEET? OR FREEZING RAIN?

Atmospheric winds, moisture content, and temperature changes going from the ground up influence if it will rain, hail, sleet, or snow. When the temperature is 32°F or below, if precipitation occurs, it will usually be snow. But if falling snowflakes encounter a sufficiently thick warm air layer, they may partially or fully melt becoming slushy drops or rain. If they descend through more freezing air, they become frozen rain drops or ice pellets, known as sleet. If you see small pellets bouncing off the ground—sleet. If rain falls into a layer of the atmosphere at or below freezing, the drops can become supercooled. This is known as freezing rain and when it hits, it can instantly encase surfaces, such as branches, wires, a windshield, or a road, in ice. The destructive nature of "insta-ice" was showcased in January 2024 in Oregon, as freezing rain toppled trees, caused accidents, and led to widespread power outages.

Hail forms very differently than sleet, within updrafts in thunderstorms. When raindrops are caught in ascending winds (updrafts) and carried upward into temperatures below 32°F, they can freeze and become hailstones. If the hailstones then fall through rain, liquid water adheres to their surface. If carried back up into colder temperatures, another layer of ice forms, and the hailstones grow. When hail becomes too large and heavy to be supported by updrafts within a storm, or if winds weaken, it falls from the sky. Weather radars can detect ice in clouds and estimate the size of hailstones that are falling from any given storm. It's a hail of a process that scientists are getting better at observing and predicting.

Sleet and hail can both cause slickened roadways and the ground to appear snow-covered. Unlike sleet, hailstones can grow dangerously large. The biggest hailstone recorded so far fell during a supercell thunderstorm in South Dakota in July 2010. It measured 8.0 inches in diameter and weighed nearly two pounds! Imagine a chunk of ice twice the size of a grapefruit hurtling down from the dark, stormy sky. Yikes! But even hail that is less than an inch in diameter can do serious damage, and when it gets above one to two inches, the consequences can be severe. Each year, hail damages crops, buildings, and cars to the tune of hundreds of millions of dollars.

WHAT IS BLACK ICE?

Clear ice atop a dark roadway or body of water is known as black ice, and it is deadly dangerous, especially on roads. The ice is clear, but below is a dark surface, which makes the ice appear black and hard to see. It is particularly dangerous in the early morning or at night when freezing temperatures have turned wet asphalt into a difficult-to-discern ice rink.

HOW MUCH IS IT GOING TO SNOW?

If snow is in the forecast, everyone wants to know: how much? But that's no easy job. There's a heck of a lot to consider when making a good snow forecast, and it can all change quickly.

As previously mentioned, temperature and moisture levels at the ground and up through different layers in the atmosphere determine whether precipitation will fall as snow, rain, freezing rain, or sleet. Sometimes, it starts to snow (or rain) but the falling precipitation encounters dry air on the way down, causing the flakes (or drops) to evaporate. Look for clouds with fuzzy bottoms, or as meteorologists call it, virga, to see this in action.

With too much dry air, an expected snowstorm may turn into just a flurry. But with a lot of moisture throughout the atmosphere, snow might fall and keep coming and coming . . . and coming. If the storm is intense enough, you might get one of many meteorologists' favorites—thundersnow! If it's relatively warm, like near 32°F, the snow could be saturated with water. When this happens, one inch of liquid (rain) can produce about three to six inches of wet, slushy snow. But if it's very cold, one inch of liquid could generate fifteen to twenty inches of dry, fluffy snow. Add sustained winds of thirty-five miles per hour for three hours or more, with visibility less than one-quarter of a mile in blowing or falling snow, and it's officially a blizzard.

If snow falls through relatively warm air, it may melt into soft frozen pellets, or graupel. This happens a lot at higher elevations. Now add changes in temperature and moisture over time throughout the atmosphere, plus changing winds, additional moisture from a lake or ocean, and mountain elevations that can block wind or warm it up. And if the ground is warm, snow may melt without accumulating. Whew! So, give your meteorologist a break if the snow forecast is uncertain or a little off; they're juggling more than snowballs to get it right.

WEATHER FORECASTING AND EXTREME EVENTS

IT SEEMS LIKE EVERYTHING IS BEING CALLED EXTREME. WHEN IT COMES TO WEATHER, WHAT IS EXTREME?

In sports, the old saying is records are made to be broken. In terms of weather, sometimes we'd prefer the same old conditions or even less so when it comes to hurricanes, droughts, flooding, and wildfires.

Based on global air-surface temperatures, 2023 was the warmest year in recorded history (since the 1940s). High temperature records were broken throughout the year across the United States and throughout the world. More than broken, they were smashed! Heat waves seared the United States, South America, Europe, China, the Middle East, and Africa. Even in the ocean, temperatures soared.

Along with heat, 2023 brought rain. Tons and tons of rain. Intense precipitation and "rare" flooding events inundated regions in the United States, India, China, Japan, Africa, and Europe. In Greece, Daniel, a hurricane-like storm in the Mediterranean, or "Medicane," dumped the equivalent of eighteen months' rain in just two days. After Greece, Daniel struck Libya. Torrential rains caused flash floods and the collapse of two dams, killing more than ten thousand people. And Hurricane Otis slammed into Acapulco. But it wasn't the only Cat 5 storm of the year. In fact, for the first time on record, every tropical basin around the world recorded at least one Cat 5 storm.

Then there were the fires. Intense, widespread, and unprecedented fires raged in Canada, creating smoke that swept into the United States, causing days of choking unhealthy air. In other places, such as Hawaii, Greece, and Spain, apocalyptic wildfires caused mass evacuations, tragic loss of life, and widespread destruction.

In 2023, according to NOAA, the United States had 28 weather/climate disaster events that resulted in losses of more than a billion dollars each. This broke the previous 2020 record number of 22 billion-dollar disasters, obliterated the 1980–2023 annual average of 8.5 events, and exceeded the most recent five-year average of 20.4 events. Across the world in 2023, there were a record 63 billion-dollar disasters. On land and in the sea the extreme heat continued in 2024 and events were no less dramatic and costly, including the deluge that struck Dubai in April with five inches of rain in just twenty-four hours. It was the highest rainfall in the seventy-five years of records and more than what the area averages in a year. The whole idea of a one-in-however-many-years storm, wildfire, or flood has gone the way of the dodo. Weather-related events once thought rare or extreme are happening more frequently and in more areas. Should they still be called extreme, or the new normal? How can anything that results in tragic loss of life and costly infrastructure destruction be considered normal or ordinary? And with carbon dioxide concentrations in the atmosphere still rising, the planet continues to warm, meaning more change is on the way—whether we like it or not. The extremes of today may be just the beginning of what conditions will be like in the years to come (more on climate change in the next chapter).

CAN EXTREME IMPACTS AND EVENTS BE PREDICTED?

During Hurricane Harvey in 2017, even before a single drop of rain had fallen, the NWS issued a bold and confident forecast for high-impact flooding three days ahead of time. The NWS forecasters at the Weather Prediction Center (WPC) were so confident in the heavy rain forecast they sought permission and received approval from headquarters to issue an unusual "high-risk" forecast. Final rainfall totals in the Houston, Texas, area exceeded a record-breaking sixty inches. Warnings were also issued as a major storm was forecast for Dubai in April of 2024. Oh, and by the way, cloud seeding had nothing to do with the intense precipitation and unprecedented flooding.

Within hours to a day or two, meteorologists can now well predict the *probability* or *chances* of extreme events, such as intense precipitation, heavy snowfall, hail, severe storms, and tornadoes. They still cannot give pinpoint locations or specific timing or say with 100 percent certainty such events will occur. But with the likelihood of severe weather, they can warn those potentially in harm's way to prepare for the impacts and stay alert.

But extreme forecasts come with many challenges. Some decision makers may take a conservative path and downplay or slightly alter the magnitude of potential impacts to avoid being wrong if such extremes don't transpire. Not everyone will receive the alert or warning, or they may not understand what the information means. During Hurricane Ian in 2022, catastrophic storm surge was forecast. But some people prepared only for the dangers of high winds. In some areas, evacuations were ordered twenty-four hours later than they could have been, and far too many people were unprepared when ten to fifteen feet of water rushed ashore in Fort Myers Beach, Florida. The consequences were tragic and prompted serious investigation into what went wrong.

Social scientists and meteorologists are working together to study how high-confidence forecasts and warnings are communicated and what messaging is most effective in getting people to take appropriate lifesaving actions. Ensuring that people are aware of extreme events and weather-related hazards, understand their risks, and know how to respond is a tremendous challenge but essential to saving lives and property.

WHAT'S THE MOST DANGEROUS WEATHER?

Before the summer of 2023, the answer to this question might have surprised most people. It's not tornadoes or even hurricanes. According to the National Weather Service, on average, over the previous ten to thirty years, the most fatalities and injuries were caused by . . . extreme heat. And the numbers are escalating quickly.

In 2022, Europe sweltered in a record-breaking heat wave that reportedly caused the heat-related deaths of more than seventy thousand people. With the scorching heat of 2023, the number of heat-related deaths is sure to have been even greater. In Phoenix, Arizona, alone, 579 lives were lost due to heat in 2023. In extreme heat, vulnerable populations are especially at risk, but without precautions, overheating and dehydration pose serious risks to everyone, even the fittest of athletes, and can be fatal. It can lead to heat exhaustion, heatstroke, heart attacks, strokes, and other serious health consequences. Adding to the danger of extreme heat is the potential for power blackouts due to heavy electrical loads. As summer temperatures continue to soar across the globe, the demand and need for electricity escalates. When critical infrastructure fails, blackouts combined with extreme heat pose an imminent threat. What about places where cooling is unavailable or unaffordable? What about workers whose jobs require exposure to high heat conditions? Extreme heat is a very real, silent, and deadly danger.

In the United States, as of today, the Federal Emergency Management Agency (FEMA) does not include heat in its response plan. In extreme heat, states cannot declare a disaster and obtain federal assistance for their response, including an all-hands approach to preventing mass casualties. As climate change ramps up the temperature, this needs to change, and soon.

Getting back to the list of most dangerous weather: Flooding (including storm surge) and rip currents consistently rank as the next most dangerous weather-related hazards, with hurricane winds, tornadoes, winter, cold, and wind in general coming in next depending on the year. Lightning is also on the list.

IN A CAR WITH THE WINDOWS CLOSED, HOW FAST DOES THE HEAT RAMP UP?

The answer may surprise some, but it's critically important to understand because far too many children and pets are lost each year as a result of overheating in cars. Even on a 61°F day, the inside of a parked car can rapidly turn deadly. If the car is in the direct sun, in just an hour, the temperature inside can soar to 105°F. According to medical experts, if a child's core temperature reaches 104°F, they are at risk of heatstroke. At 107°F, the consequences become fatal. Today, summer heat is more extreme than ever, and hot temperatures are lasting longer into the fall. With daytime temperatures regularly over 90°F or even 100°F, a locked car becomes a death trap frighteningly fast. *Always* make sure you look in the back seat if you have children or pets.

Another heat-related caution: Please don't walk your pet when asphalt temperatures are scalding hot. Believe it or not, an air temperature of just 77°F can, in the sun, result in asphalt temperatures around 125°F. That's much too hot for your four-legged family members. Remember, if the asphalt is too hot for the back of your hand, it's too hot for your pet's paws as well.

CAN I OR SHOULD I DRIVE THROUGH THAT WATER IN THE ROAD?

We know that trucks are tough. Heck, a major truck manufacturer has built its reputation on building tough trucks. But even tough trucks aren't battle-tested against flowing water. It is an astonishingly powerful force. Whether it's storm surge, a flooded roadway, or an overflowing creek, a foot and a half of flowing water will sweep a person off their feet. Just a little more, and cars go afloat. Here's the thing: It is impossible to discern the depth when driving into a puddle or a flooded roadway. Or to see if the underlying road has given way. Don't take a chance. If water is flowing over the road or there's standing water pooled

Turn around and don't drown. No matter what size your vehicle is.

ahead, turn around and don't drown. And for the love of God, if a barrier is blocking the road, do not go around it!

IF CAUGHT ON THE ROAD IN A TORNADO, WHAT SHOULD I DO?

With its flying cows, special effects, and characters passionate about science, *Twister* is one of our favorite movies. It was also the spark that inspired many a new meteorologist. But bloody hell, if a tornado is headed your way, do not go into a barn stocked with all manner of hanging saws and blades (sheds, storage facilities, and mobile homes are not safe shelters). Do not go under a bridge overpass and grab onto a support for protection (narrow open areas will increase the windspeed). And no, there will be no mooing cows (or sharks) flying out of a tornado.

Meteorologists can now forecast days in advance the likelihood that powerful storms may produce tornadoes. With radar, they can identify patterns that indicate rotation in the atmosphere, which may indicate a tornado is developing. Radar can also pick up debris from destructive tornadoes on the ground. However, tornadoes can spin up rapidly, appear quickly, and change suddenly in size or direction; warnings may be short at best. It is critical to be weather aware if severe storms are forecast for your area. If a tornado warning is issued, seek safe shelter ASAP, such as underground in a cellar or basement or in a small interior room (such as a bathroom) well away from windows. Take your pets with you, and do not go out to take photos or video when a tornado is close.

There's new hope on the horizon for improved warnings for tornadoes, severe thunderstorms, and flash floods. NOAA has been working on its experimental Warn on Forecast technology to provide more advanced warnings. For tornadoes, they're hoping for lead times up to sixty minutes. If it works, this will give people additional time to seek safe shelter. Results are already promising.

To be clear, a car is not a safe place to be. Get out and seek safe shelter. If a building is not available, find the lowest ditch possible (as long as it is not flooded or flooding) and cover your head. If you must stay in your car, get down as low as possible and again, cover your head.

It is always best to have a plan for where to go should a tornado warning or alert be issued for your area. If a tornado watch is issued, think hard about any travel plans that day that necessitate driving, and whether they are truly necessary. The bottom line: You do not want to be in a car or stuck in traffic if a tornado is anywhere nearby. Being weather aware on these days could save your own life and the lives of others.

CAN I OUTRUN A TORNADO?

No!

WHAT ABOUT RIP CURRENTS? CAN I SEE AND AVOID THEM?

Most of the time, no. But occasionally, lines of sediment or debris or an area of low wave breaking may mark a rip current—a relatively narrow, high-velocity jet of seawater flowing offshore.

Each year about a hundred people die in rip currents. Sadly, we include a beloved colleague in this statistic. Fatalities occur due to exhaustion, panic, or poor swimming skills. Misconceptions about rip currents include that they occur only in rough or high-wave conditions (nope), can be seen from the beach (occasionally), and pull people under (that's an undertow).

Rip currents may be present even when waves seem small and winds light. They form when seawater piles up on a beach (due to onshore winds, tides, and storms) and then flows back to the sea. Low areas onshore or obstructions such as a jetty or seawall can funnel the flow seaward, creating or enhancing a narrow, fast current.

Not even Olympic swimmers can successfully battle a rip current, which according to NOAA can flow up to eight feet per second. If caught in a rip current, it's best to relax, float, and call for help by waving your arms and yelling. If no lifeguards are present, swim parallel to shore until free of the rip current and then head toward the beach. Never try to swim back to shore directly against the flow—that's a surefire recipe for exhaustion and drowning.

The National Weather Service now provides forecasts of the potential for rip currents, and lifeguard stands often have warning flags up if rip currents are present. It is always safest to swim with a buddy, where lifeguards are on duty, and look for warning signs and flags before entering the water. Know before you go!

CAN I OUTRUN A TSUNAMI?

To be clear, tsunamis are not weather-related events. But meteorologists are often called upon to discuss and provide information about tsunamis. As for outrunning one—that, too, is a big no.

Tsunamis are generated by large displacements of water. They can be caused by vertical movements of the seafloor (earthquakes), landslides, volcanic eruptions, and asteroid impacts. The energy from the triggering event is transformed into wave motion that travels across the ocean. In the deep ocean, the waves are long, low, and fast—traveling at about five hundred miles per hour. As they approach the shore and shallower water, tsunami waves begin to "feel bottom." Friction causes the waves to slow at their base, while the top speeds ahead. The waves steepen, grow in height, and may break. Typically, when striking the shore, tsunamis can slow to about twenty miles per hour, which is still too fast for a foot race.

Tsunamis come ashore as a series of waves or surges, like a superfast incoming tide. The height, shape, and number of waves depends on the magnitude of the triggering event, the bathymetry (depth), and the nature and shape of the coastline. Coral reefs, mangroves, and wetlands can help protect the shoreline and dampen the wave energy. Powerful surges onshore can also create swift, debris-laden, and dangerous currents that flow back to the sea.

If a tsunami warning or alert is issued, or a strong tremor that lasts more than thirty seconds is felt at the coast, head to high ground, and fast. Do not wait to see what happens. And if the sea recedes quickly, like a rapid outgoing tide, do not go out to investigate, record video, or pick up flapping fish. Get to high ground as soon as possible.

WHAT IS A METEOTSUNAMI?

When abnormally high waves, up to six feet high, strike the shore out of the blue, it can cause flooding, injury, and damage to roadways and buildings. Scientists now recognize some of these events as meteotsunamis—tsunamis that are generated by a meteorological event like a fast-moving storm or squall. Rapid changes in air pressure can generate long, low, fast-moving waves in the open ocean. When the waves reach the shore, shallowing and geography can magnify their size and increase the severity of impacts. Meteotsunamis have occurred along the U.S. East and Gulf Coasts, in the Great Lakes, and in the Mediterranean Sea and are very difficult to predict. NOAA estimates that about twenty-five meteotsunamis occur each year along the U.S. East Coast, but luckily most are under a foot and a half high.

■ ■ ■

When it comes to weather, be aware and be smart. We provide our sources at the back of the book and encourage you to explore these topics in more detail.

Chapter Ten

CLIMATE CHANGE

A METEOROLOGIST and an ocean scientist walk into a bar in Maine and order margaritas (true story). They joke with the bartender and converse with two men sitting nearby. Weather is soon a hot topic, and they are asked: "Hey, what about this climate change thing? Is it real?"

The meteorologist gestures for another, stronger margarita. The marine scientist says make it two, and they proceed to respectfully answer the question, being sure to relate climate change to what the men care about (fishing, coastal real estate, food, and quality of life). The conversation is informal and light, and by the end the men are receptive to the idea that Earth is warming at an unnaturally accelerated pace, impacts are on the rise, and humans are to blame.

Given the destructive, in-your-face extreme events across the globe in the past few years—record-breaking scorching heat waves, out-of-control and tragic wildfires, unprecedented and catastrophic flooding events, and exceptionally powerful storms—we hope fewer people are questioning the reality of climate change. But we've been working with leading scientists and broadcast meteorologists to communicate accurate information about climate change for years. We've been immersed in the science involved, tracking the data, reading articles, leading weather and climate summits, and watching climate-related events wreak havoc worldwide. For most people, we know this is not the norm. Some people remain unsure or have questions, doubts, and perhaps misunderstandings about climate change. So, here are some of the questions we and our colleagues regularly get about climate change. In our answers, we try to keep the jargon to a minimum and use easy-to-understand language. What we present is based on scientific evidence and data, details of which can be found in the chapter references and sources at the back of the book.

WHAT'S THE DIFFERENCE BETWEEN WEATHER AND CLIMATE?

Weather is what we experience day to day, and climate is what occurs over the longer term. In other words, climate is what we expect, and weather is what we get. According to NOAA's National Centers for Environmental Information (NCEI), when assessing climate, a thirty-year record is a good standard and a sufficient time frame to use.

Here's another way to look at the difference between weather and climate. If it's hot in the summer, you might think it's just summer weather. Summers are hot. But how does this summer

CLIMATE CHANGE

In rising heat, wildlife may search for cooler stomping grounds.

compare to the previous thirty summers? If you take the average of the daily temperatures over time for your location, is it getting cooler or warming up? What's the long-term trend? That's climate.

ARE GLOBAL WARMING AND CLIMATE CHANGE THE SAME THING?

It seems confusing when everyone used to call it global warming and now it's climate change. But there's a real difference. Global warming typically refers to the rise in average surface temperatures around the world over time. Climate change is a more comprehensive description of what's going on. In addition to the increase in global temperatures, it includes associated impacts such as sea-level rise, increasing acidity in the ocean, the melting of glaciers and ice caps, permafrost thaw, extreme weather events, wildfires, algal blooms, decreasing oxygen concentrations in the ocean, coral-reef bleaching and mortality, wildlife migrations, and changes in ocean circulation. Climate change also includes the impacts on society in areas such as agriculture, food security, transportation, public health, safety and security, the tourism industry, human migrations, military operations, and conflicts. Global warming is one part of the entirety that is climate change.

HOW DOES CARBON DIOXIDE CAUSE WARMING?

Hundreds of years ago, scientists recognized how carbon dioxide in Earth's thin atmosphere helps to warm the planet and that without it and other so-called greenhouse gases (water vapor, methane, nitrous oxide, and chlorofluorocarbons), the planet would be more like Mars, and at night, frigidly cold. Here's how it works.

Incoming shortwave energy or radiation from the Sun passes through Earth's atmosphere and strikes the surface. Some energy is reflected, especially from light-colored surfaces like ice or snow. Much, however, is absorbed and then reemitted as longer-wave radiation. Gases in the atmosphere, principally carbon dioxide, water vapor, and methane, absorb this energy and reemit it as heat, warming the planet. The more carbon dioxide in the atmosphere, the more energy that can be absorbed and reemitted. It's like piling the blankets on the bed at night.

Recognizing this process, scientists also realized hundreds of years ago that with increasing concentrations of carbon dioxide in the atmosphere, Earth's surface temperatures would warm.

THERE'S NOT VERY MUCH CARBON DIOXIDE IN THE ATMOSPHERE. HOW COULD IT POSSIBLY MAKE A DIFFERENCE?

Approximately 99 percent of Earth's atmosphere is nitrogen and oxygen. Carbon dioxide makes up only about 0.04 percent. But unlike nitrogen and oxygen, carbon dioxide is a very potent greenhouse gas, meaning it readily absorbs the long-wave radiation emitted by Earth. Things that are potent tend to be exceptionally strong in small quantities. A fatal dose of arsenic for an adult is just 0.00032 pounds. An even more miniscule amount of fentanyl is deadly. Just because something is present in small quantities doesn't mean it is inconsequential or cannot have big impacts.

IN EARTH'S PAST, HAVEN'T CARBON DIOXIDE CONCENTRATIONS BEEN HIGHER THAN THEY ARE TODAY?

In 1958, Charles David Keeling of Scripps Institution of Oceanography began making direct measurements of atmospheric carbon dioxide atop Mauna Loa, Hawaii. The Mauna Loa Observatory, run by NOAA, now hosts the longest-running record of the carbon dioxide concentration in Earth's atmosphere. Since the 1950s, it shows that carbon dioxide levels have gone from around 300 parts per million (ppm) to about 425 ppm.

Yes, in the past, carbon dioxide concentrations in Earth's atmosphere have been as high as or higher than they are today. But—and this is a very big but—the planet was hotter, more ocean-covered, and much less hospitable for its residents.

Based on scientific data, about fifty million years ago, atmospheric concentrations of carbon dioxide were around 1,000 ppm. There were no ice sheets or glaciers, global temperatures were 14°F to 21°F warmer than today, and sea level was more than two hundred feet higher. Approximately three million years ago, atmospheric carbon dioxide concentrations were about what they are today (350 to 400 ppm), Earth was about 2°F to 5°F warmer, and sea level was around sixty-five feet higher. And just 125,000 years ago, when carbon dioxide concentrations were about 300 ppm, the planet was again 2°F to 4°F warmer and sea level was about eighteen feet higher than today. It's not the planet we're worried about. But imagine extremes even worse than today: greater and more widespread life-threatening heat, more rapidly spreading and intense wildfires, stronger storms, increased flooding, and—here's the big one—sea level eighteen feet or more higher than today.

Most of the excess heat on Earth today, at least 90 percent, has been absorbed by the ocean. Were the sharp rise of ocean temperatures and increasing heat content in 2023 telltale signs that the ocean and planet are beginning to adjust?

Will it lead to conditions more like those 125,000 or three million years ago? It's hard to accept that we are headed in that direction.

CAN'T WE JUST ADAPT TO THE CHANGES?

As temperatures rise on the planet, can't we just make the changes needed to live comfortably? Yes, but not fast enough. Unfortunately, today's warming is occurring at too fast a pace and is rapidly creating conditions that were not considered when planning critical infrastructure, water and food supplies, community development, industry, disaster response, and more. Our societal foundations were built for climate conditions that no longer exist. It will take time to do things such as build heat-resistant, energy-efficient cities; move coastal communities or adapt them to rising seas and extreme storms; develop food sources adapted to rising temperatures, droughts, and flooding; and create safe, sustainable water supplies. As resources diminish and conditions become more extreme, some regions will become unlivable, creating conflicts and refugee populations. In some areas of Africa, climate-related disasters are already creating large refugee populations. Simply adapting isn't an option.

And it's not just us. In the past, when warming occurred at a slower rate, animals and plants had time to adapt. Today, the pace of warming is too fast for many species to acclimate. Some species are migrating to regions with more amenable conditions, but for those unable to do so, populations are decreasing, and extinctions are on the rise. Overall, the diversity and abundance of species on the planet is at risk. How this will impact the livability of Earth for humans is at best poorly understood.

HOW DO YOU KNOW WHAT THE TEMPERATURES WERE A THOUSAND YEARS AGO WHEN THERE WERE NO THERMOMETERS?

Meteorologist Mike Bettes of the Weather Channel says this is a question he hears frequently. It's true that thermometers weren't around a thousand years ago. What about temperatures quoted for 100,000 or more than a million years ago? Where do those come from?

Climate change expert Dr. Jim White, a professor and dean at the University of North Carolina, Chapel Hill, likes to explain that many physical things change as temperatures change, and while these changes may not provide the accuracy of modern thermometers, they can be used to distinguish hot from cold or warming from cooling. For example, trees rings record growth that is dependent on temperature and precipitation. A record of temperature during growth is also preserved in the skeletons of corals and the shells of small marine animals called foraminifera. Fossils or pollen within undisturbed layers of sediment in lakes and oceans offer another means to decipher past temperatures. Ancient temperatures as well as carbon dioxide can also be studied by drilling cores through old glaciers to analyze air trapped within bubbles in their thick, compact layers of ice.

Scientists can now combine these types of proxy data with modern observations to reconstruct global temperatures and carbon dioxide concentrations going back hundreds of millions of years.

THE CLIMATE IS ALWAYS CHANGING. ISN'T THIS JUST A NATURAL CYCLE?

Through reconstructions of Earth's past temperatures and carbon dioxide concentrations in the atmosphere, we know that, yes, Earth does go through natural climate cycles. For example, based on data from Antarctic ice cores, over the past 800,000 years and up until the 1950s, Earth's temperatures went up and down and carbon dioxide concentrations in the atmosphere varied between about 170 and 300 ppm.

Data also indicate that in the past, such planetary swings from warm to cold and back were driven by massive volcanic emissions, the distribution of land and sea, ocean circulation, tectonic upheavals and weathering, and orbital variations (Earth's orbit about the Sun, the tilt of the planet's rotational axis, and the wobble of the planet's spin). Carbon dioxide played a role, but rather than driving change, it enhanced or reduced the effects of the other factors. But that's not what's happening today. We can use well-tested computer models and observational data to show that increasing carbon dioxide in the atmosphere is now overshadowing the other factors and has become the main driver of climate change.

For instance, when only natural factors, such as the Sun's energy output, volcanic eruptions, and Earth's orbital cycles, are included in climate models, today's rising temperatures cannot be reproduced. But when increasing concentrations of carbon dioxide in the atmosphere due to human activities are included, the models reproduce the rise of global temperatures. (More to come on climate models.)

Scientists have also been able to determine the origin of the increasing carbon dioxide in the atmosphere. Stay with us here, as it gets a little complicated. Carbon has three naturally occurring isotopes (atoms with a varying number of neutrons and different atomic weights): light carbon-12, heavy carbon-13, and radioactive carbon-14. During photosynthesis, plants use carbon dioxide from the atmosphere and preferentially take up the lighter isotope of carbon (carbon-12). When plants (young or old) are burned, the lighter carbon is released back into the atmosphere. Since the Industrial Revolution, measurements of carbon dioxide in the atmosphere show that the ratio of heavy (carbon-13) to light carbon (carbon-12) has decreased—the amount of light carbon released from plants has risen. This is supported by data on emissions and deforestation.

Research led by atmospheric scientist Dr. Ben Santer at the Lawrence Livermore National Laboratory on long-term temperature trends in the upper stratosphere also shows a clear human imprint on the changes now occurring. Results published in 2023 conclude "it is now virtually impossible for natural causes to explain satellite-measured trends in the thermal structure of Earth's atmosphere."

Bottom line is that we can fingerprint the excess carbon dioxide and temperature changes in the atmosphere: They are principally the result of the burning of ancient plants, aka fossil fuels, such as coal and oil.

WHAT ABOUT VOLCANOES?

Nope. There's no evidence that an increase in volcanic emissions has generated the modern rise in atmospheric carbon dioxide or temperature. This includes the 2022 Tonga eruption that put massive amounts of water vapor into the atmosphere and may temporarily contribute to a small part of current warming.

WHAT ABOUT THE SUN?

Nope. There's also no evidence that an increase in solar radiation (including sunspot activity) has occurred that can explain Earth's rising temperatures (see next chapter for more).

WHAT ABOUT RECENT REDUCTIONS IN SHIP EMISSIONS?

New regulations requiring the shipping industry to use cleaner fuels have reduced sulfur emissions. These emissions can have a cooling effect on the climate by reflecting incoming sunlight and helping to create clouds. Have reduced emissions driven or contributed to today's warming? Data suggest a negligible impact on global mean temperatures, on the order of a few hundredths of a degree. So, no, it's not responsible for the current extreme and rising heat.

CLIMATE CHANGE

WE ALWAYS HEAR ABOUT THE MODELS. CAN WE TRUST THEM?

Computer models are reliably used every day: for short-term weather forecasts, to track and forecast hurricanes, in flight simulators to train pilots, to predict disease outbreaks and spread, in traffic control, and for spectacular computer-generated imagery (CGI) in the movies. Computer-based or numerical climate models incorporate our understanding of the physics and dynamics of the atmosphere and the ocean, the distribution of land and sea on the planet, ice cover, solar radiation, volcanic emissions, and an array of observations as input. Unlike weather models, climate models examine long-term trends. But are they any good?

To determine their reliability, climate models can be tested in several ways. Data on conditions in the past can be input and the models run to see if they reproduce previous climate conditions. They do. The way we test computer models for today's climate is to compare the results with real-world observations. Research on the accuracy of global climate models over the past half-century has shown that they are quite accurate. Large-scale features like the jet stream are well represented, and resulting global temperature changes are generally in agreement with observations. Even the stunning rise in temperatures in 2023 fall within the range of predicted warming. Global climate models, however, cannot simulate specific, small-scale weather events such as intense storms or flood events.

Using current data such as solar radiation, volcanic emissions, and orbital variations, the models can also be used to test what has the most influence. The models clearly show that today, unlike in the past, the rising level of carbon dioxide in the atmosphere is the dominant driver of Earth's escalating temperatures.

IT'S SNOWPOCALYPSE. WHERE'S THAT GLOBAL WARMING NOW?

In 2022/2023, snow fell not in inches but in tens of feet in the western United States, burying previous records. In 2010, when up to three feet of snow buried Washington, DC, from back-to-back storms, it was dubbed Snowmaggeddon. Events such as these, or just a cold snap, get people asking: Where did that global warming go now?

In a warming world, heavier snowfalls and more intense rains are expected. For every degree Fahrenheit of warming, the atmosphere holds 4 percent more water. More water in the atmosphere means that when it rains, it rains more intensely. When it snows, it snows harder. Unfortunately, it doesn't mean additional rain or snow spread out over time or area; it means more extreme events like record-breaking downpours and unprecedented snow accumulations. Contributing to these extremes are changes in the jet stream, which appears to be slowing and becoming wavier, leading to stalled and longer-lasting storms (as well as heat waves). When you hear about "heat domes" and "cutoff lows," they refer to low-pressure systems that have been cut off from the jet stream. Even as average global surface temperatures continue to rise, we can still expect some seriously snowy and frigid days . . . but perhaps not as many.

CLIMATE CHANGE

BUT IN THE 1970S, DIDN'T SCIENTISTS SAY THE WORLD WAS COOLING?

Not exactly. In the early years of "climate science," a handful of researchers suggested that cooling was imminent. A myth later arose that a consensus or majority of scientists believed temperatures would swing downward. Nope, but the media at the time fueled the "cooling myth" with some selective misreading and/or misunderstanding of the science. In more recent years, the purported cooling idea has been promoted by nonscientists, often with an agenda, such as celebrity novelist Michael Crichton. In his 2004 novel *State of Fear*, a character claims, "Back in the 1970s, all the climate scientists believed an ice age was coming." The author later suggested that global warming was just a ploy to create a crisis necessitating funding. A 2008 review of the 1970s literature, published in the *Bulletin of the American Meteorological Society*, showed that imminent cooling was in no way a consensus among scientists and that, even then, the scientific literature was dominated by concern over warming caused by greenhouse gases. Oh, and that purported 1970s *Time* magazine cover showing a penguin, suggesting that an ice age was coming? It was a *fake*—doctored, aka not real. The original 2007 cover had the headline "The Global Warming Survival Guide."

ISN'T MORE CARBON DIOXIDE GOOD FOR PLANTS?

We completely understand the thinking here. Plants use carbon dioxide for photosynthesis, so more must be better. And some plants do seem to be doing better . . . like (uh-oh) more potent, fast-growing poison ivy. For pollen allergy sufferers, it's definitely not good news. Pollen-producing plants growing better mean more pollen and more intense or longer allergy seasons. But plant growth also depends on other factors, such as temperature, water, soil, and nutrients. Extreme heat, too much or too little rain, washed-out or extremely dry soil, and intense wildfires are typically bad for plants. Because more carbon dioxide is also resulting in conditions and events that negatively affect plant growth, it's not good for them either.

HOW FAST AND HOW MUCH WILL SEA LEVEL RISE?

This is a big question with frustratingly big uncertainty. Remember, no one was here last time Earth's glaciers and ice sheets melted like they are today. What we do know is that as the planet warms, global sea level rises because of the thermal expansion of seawater and the melting of land-based glaciers and polar ice caps. The melting of floating ice shelves or sea ice does not contribute to sea-level rise—just as when ice cubes melt in a drink, the liquid doesn't overflow the glass. Local and regional sea-level rise is also influenced by the sinking or subsidence of land, tectonic events, wetland loss, groundwater depletion, gas and oil extraction, ocean circulation, and the rebounding of land from ice loss or weathering.

Since 1880, the average global sea level has risen more than eight to nine inches. The annual rate of sea-level rise accelerated in 1993, from about .08 inches to .11 inches. This seems like a small amount, but over time it adds up. Over thirty years, sea level has risen three inches. Add another thirty years at that rate, and the ocean will be half a foot higher. If the pace is faster or accelerates (which it is doing), sea level will rise even more and sooner. Data indicate that the accelerated pace of global sea-level rise is due mainly to the melting of land-based ice in Greenland and Antarctica.

In some areas, sea level is rising even more rapidly. Between 2010 and 2022, along the U.S. southeastern and Gulf coasts, sea level rose an astonishing 0.4 inches per year, or nearly five inches . . . getting close to that half-a-foot mark. The increased rate of sea-level rise in these areas may be related to periods when circulation in the Atlantic Ocean slowed, and subsidence of the ground.

Even relatively small rises in sea level can have big impacts in low-lying areas and already-flood-prone regions. In the United States, increased tidal flooding is already occurring in many states, including Texas, Florida, New York, New Jersey, Maryland, and Virginia (plates 16 and 17). Sea-level rise also increases storm surge, exacerbates coastal erosion, and causes saltwater intrusion. In New Orleans, Louisiana, drought combined with sea-level rise is allowing saltwater to propagate up the Mississippi River (saltwater intrusion), threatening to contaminate drinking water supplies for tens of thousands of people or more. With a rising ocean, roadways are flooded more frequently, bridge or building foundations can be weakened, and crops are lost. Other impacts include escalating insurance rates, loss of real estate value, and the need to retreat from coastal living spaces. Many island, coastal, and even inland communities around the world are already experiencing the costly impacts of sea-level rise.

As the planet continues to warm, the sea will continue to rise. Using the best models, observations, and data available, estimates suggest that in the United States, sea levels will rise about one foot over the next thirty years—as much as it's risen in the past one hundred years.

But the ocean has risen only a few inches, and already it's causing major problems. Imagine what one foot will do. Coastal highways will be submerged, and more homes will be unlivable if not destroyed. More businesses will be lost, more drinking water contaminated, more crops lost, and entire communities forced to relocate. And it's not just small coastal towns that are and will be affected. Remember the flooding in New York during Hurricane Sandy? A 2019 projection for Miami–Dade County, Florida, suggests that by 2040 sea level may rise ten to seventeen inches!

Could the ocean go up even faster? Scientists who study Earth's disappearing ice are concerned that predicted estimates for sea-level rise are too conservative. The ocean may go up faster and more than expected. How much faster, how much higher, and when? Scientists around the world are making observations, collecting data, running models, and working hard to answer that question.

ARE OCEAN CURRENTS SLOWING?

In the climate-disaster movie *The Day After Tomorrow*, ocean circulation played a starring role. With the film came heightened interest in the ocean's flow and whether changes could produce phenomena like killer tornadoes in Los Angeles and instafreeze land-based hurricanes. Probably not these last two, but changes in ocean circulation do influence Earth's climate and weather.

Ocean circulation helps to distribute and balance heat on the planet and affects regional climates. Changes in temperature and salinity (density) drive the ocean's large-scale circulation, which in a simplistic way flows around the globe like a giant conveyor belt. Warm waters tend to flow on the surface and cold, salty waters flow at depth. Wind, tides, gravity, mixing, and geography also influence flows within the sea.

In the context of climate change, a commonly asked question in the science world is: *Is the increasing input of meltwater in the Arctic and/or Antarctic causing the ocean's large-scale density-driven circulation to slow?* Geologic evidence suggests that when this happened in Earth's past, the climate changed abruptly.

The thinking here is that if large amounts of freshwater are released into the ocean at high northern or southern latitudes, it could reduce or stop the formation of cold, salty deepwater. If this happens, it could trigger a global ocean traffic jam. Without the formation (due to evaporation and chilling) and push of cold, salty deepwater, other flows would stall or be blocked. For example, it could impede the northward flow of warm surface water in the Atlantic—the Gulf Stream. Less heat would be transported north, and northern Europe would cool dramatically. The southeastern coast of the United States would warm and experience higher sea levels (due to the thermal expansion of seawater).

Observations suggest that in the modern ocean, global circulation has gone through periods of slowing and then restrengthening. In 2009–2010 a slowdown occurred, and the ocean's great flows turned sluggish. The following winter in the United Kingdom was especially snowy and the coldest since 1890. At the same time, along the U.S. southeastern and Gulf coasts, sea level rose almost five inches.

Reports suggest that over the past decade, global ocean circulation has weakened. But there are a lot of unknowns and complicating factors. We're not even sure where meltwater from Greenland goes or what other influences may be at play, including atmospheric dynamics, variations in El Niño–La Niña, the shifting position of tropical moisture and winds, and the influence of sea-ice stability as well as high-elevation ice sheets. What about the unprecedented marine heat waves and loss of Antarctic sea ice in 2023? Are these conditions the result of the ocean and planet adjusting to rising warmth? Is it a consequence of a slowing ocean, or will it cause more slowing? The unwanted answers to these questions may be coming soon.

EXTREME EVENTS

Sometimes words just aren't enough. As we started to write this section, the town of Lahaina on Maui, Hawaii, was annihilated by a nearly unthinkable and catastrophic wildfire fueled by high winds, invasive grasses, and climate-driven drought. The official death toll was near one hundred souls lost. Tens of thousands were left homeless, without jobs and businesses, and an entire community was burned to the ground. But Maui's residents are not alone in their pain or suffering. For millions of people, the summer of 2023, the warmest in recorded history in the

Northern Hemisphere, brought devastating climate-related events. As noted in the previous chapter, calamitous wildfires, disastrous flooding, and unprecedented heat waves struck across the globe. Extreme drought brought mighty rivers to their knees, disrupting transportation and supply chains. In multiple locations, temperatures soared to levels once thought impossible, exceeding 120°F.

Extreme events driven by climate change are not a thing of the future, a predication yet to come. They are here and happening now, causing immense, tragic, and costly destruction. While climate change doesn't create a hurricane, start a wildfire, or cause a rainstorm, it makes these events more likely and extreme. Drought and heat prime the land for larger wildfires, and their burn scars set the stage for mud and debris flows. More intense and lingering rainfall cause disastrous floods. Greater ocean heat content fuels stronger hurricanes that are more likely to rapidly intensify prior to landfall. Higher seawater temperatures also cause mass coral mortality. Sea-level rise worsens storm surge, increases coastal erosion, and undermines critical infrastructure. The costly and tragic impacts of extremes driven by climate change are only going to become more frequent and more catastrophic. Yup, it's bad. But it's also not too late to change the path we're on. More on this coming up.

With more media coverage today than ever before, a related question is: *How do we know these events are unprecedented?* The answer is observations. We've had good meteorological observations in many areas going back at least a hundred years, with anecdotal records for even longer. Based on previous observations, we can determine if something has ever happened

Extreme heat is not normal, even in summer.

before in recorded history. Take, for example, Miami's 2023 record-breaking summer heat wave.

Meteorologist Dr. Brian McNoldy at the University of Miami's Rosenstiel School of Marine, Atmosphere, and Earth Sciences closely followed the soaring 2023 summer temperatures and heat indices. So many records were being broken, he had a hard time keeping track. He was astonished that records weren't broken by just a few degrees, they were smashed by huge amounts that previously seemed unimaginable: "From mid-June through August 9th, Miami set 32 new daily heat index records, 15 new dewpoint records, and 12 new temperature records. Through August 9, the heat index spent 120 hours at or above 105°F. For perspective, that threshold was almost never crossed prior to the 1980s, it peaked at 49 hours in 2020, so the 120 hours in early August is striking because it's so far beyond any other year AND there was still more to come."

Unfortunately, many locations that are vulnerable to the extreme events wrought by climate change are also highly desirable living spaces. Consider retirees lured to the warmth and open spaces of Phoenix, Arizona, or coastal living in subtropical Florida. Did they plan for extreme heat emergencies or bailing insurance companies due to storm and flood risks? Riverbeds and creeks make for wonderful places to live, fish, and explore, but they are also vulnerable to once unthinkable flooding (as already experienced in Vermont, New York, Massachusetts, Texas, California, Florida, Nevada, Kentucky, and elsewhere). Floodplains offer fertile agricultural lands but are susceptible to inundation from more intense rainfall. The list of impacts made extreme by climate change is long and growing, and no doubt unwelcome surprises are just around the corner.

IS IT TOO LATE TO STOP CLIMATE CHANGE?

Climate change is happening here and now. But it's not too late to slow the pace of warming and reverse course. Now is not the time to give up, give in to denial or depression. Now is the time to act, to use existing and growing knowledge and technology to affect change. It won't happen overnight, but as we transition to renewable energies like solar and wind, we need to halt deforestation, stop methane from leaking into our atmosphere, improve energy efficiencies, build new resilient and efficient infrastructure, move to sustainable and more efficient agriculture practices and transportation, protect and restore ocean habitats, and reduce food wastes. People are already coming up with exciting solutions and taking action to make innovations happen. We just need to do more, invest more and do it sooner rather than later. In the long run, inaction will be more costly than making change now. Industry, communities, elected officials at all levels, philanthropists, non-profits, government agencies, and individuals are all part of the solution. Because if we reduce emissions, transition as much as possible to renewable energy, and lower carbon dioxide levels in the atmosphere, science suggests the planet will stop warming at its current rate and be on a path toward a cooler, less extreme future.

WHAT CAN I DO TO HELP?

Every individual can make a difference. Don't let anyone tell you differently. Maybe most importantly, you can vote for elected officials (local, regional, and national) who support investments, education, and policies to combat and reduce climate change, including transitioning to renewable energies and promoting conservation. Here's a question from us: Why are taxpayer funds being used to subsidize big oil, when they are raking in the bucks? In the Unites States, the fossil fuel industry annually receives some twenty billion dollars in government subsidies (our taxpayer dollars). In 2022, across the world, the amount big oil received in subsidies soared to seven trillion dollars.

By talking to your peers, family, friends, coworkers, or people with similar interests and religious or political views, you can help them understand climate change and combat misunderstandings and the spread of misinformation. You can also reduce your own carbon footprint where possible through efforts such as installing solar panels, increasing your home's energy efficiency, driving less or getting a hybrid or electric vehicle, reducing food waste, planting trees (not a solution all on its own), promoting sustainable agricultural practices, and eating less meat. And you can help by supporting industries, organizations, and individuals working to stop climate change and increase sustainability. We know how to fix this; we need the will and the investment to do it!

WHERE CAN I FIND RELIABLE INFORMATION?

Our friend and climate guru Dr. Jim White is frequently asked this question. In a world inundated with information from innumerable sources, how does a person know where to find credible information about climate change based on science? Dr. White typically recommends the summaries of the Intergovernmental Panel on Climate Change (IPCC) or, for people in the United States, government agencies such as NOAA or NASA that provide clear, understandable, and bias-free information. These sources provide information based on research, data collection and analysis, modeling, and observations—scientific evidence.

Look for website articles, blogs, social media posts, books, and presentations that are based on and link to credible scientific data and results in peer-reviewed research. Opinions on climate change from climate experts, scientists who have spent years studying the science of and data on climate change, are also good. But an expert in an unrelated field or a pundit who expresses opinions or beliefs based on something other than credible science should not be considered a trusted source on climate change (more in chapters 12 and 13). We encourage people to read, discuss, and ask questions about climate change. But please consider where information comes from and what it is based on.

■ ■ ■

At the back of the book, we provide our sources and some excellent references on climate change.

Chapter Eleven

THE SUN

WHY WAIT UNTIL DARK to wish upon a star when one shines super-bright during the day? That's right. The Sun is a 4.5-billion-year-old star. It's also a giant glowing ball of hot gas, mostly hydrogen and helium, held together by its own gravity. It's so big that, according to NASA, it would take 1.3 million Earths to fill it. It doesn't look that huge because it's so far away, some ninety-three million miles away. Its gravity also holds the solar system together; without the Sun, there would be no Earth or life as we know it. But if you want to wish upon our only daytime star, one essential reminder: Never look directly at the Sun without proper eye protection. Solar filters, like those in approved solar eclipse glasses, are a thousand times darker than sunglasses, blocking all infrared and UV light and most visible light (more on solar eclipses to come).

The science of studying the Sun and its influence throughout the solar system is called heliophysics. If you think that

THE SUN

> TWINKLE TWINKLE BIG STAR
> HOW I WONDER WHAT YOU ARE
> UP ABOVE THE WORLD SO HIGH
> A VERY BRIGHT LIGHT
> IN THE SKY

Our only daytime star . . . the Sun.

sounds too sciency, dull, or esoteric, here's how Dr. Michael Kirk, a NASA astrophysicist, describes his fascination with heliophysics: "The Sun is always moving. It doesn't matter how far you zoom in on the solar surface, there is always something evolving, creating an ever-changing spectacle. Gazing into that roiling sea of solar plasma, we can see magnetic loops that twist and dance, shaping the flow of charged particles that shimmer with a hypnotic beauty all their own. However, the Sun isn't just mesmerizing, it can have real impacts on our modern life. Solar storms have caused major disruptions to our technology and communications. Solar eruptions stretch far out into space and leave an indelible mark on the solar system."

Lead NASA program scientist Dr. Madhulika (Lika) Guhathakurta, one of the inventors of the word *heliophysics*, puts it another way: "The Sun is the biggest frigging object in our solar system, and we wouldn't exist without it. What else would you study? Size (gravity) matters."

We spoke with Dr. Guhathakurta, Dr. Kirk, and other leading heliophysics experts and dove into the science to address some of the most frequently asked questions and misunderstandings about the Sun and to discover more about the world's greatest star power.

A "BURNING" BALL OF GAS?

Descriptions of the Sun (and definitions of a star) sometimes include the word *burning*, but where there's no smoke, there's no fire. The Sun's energy does not come from combustion, aka burning; it is fueled by nuclear fusion within its core. There, temperature and pressure are extraordinarily high, which causes the protons in hydrogen atoms to smash into one another and fuse to become helium atoms. During this process, some mass is converted into energy, released, and radiated outward to the Sun's surface and into its outer atmosphere, or corona. From there, it is emitted as energy, or electromagnetic radiation—some of which zooms toward Earth at the speed of light, making the trip in just over eight minutes. Most of the Sun's radiation reaching Earth is visible light and invisible infrared energy; a small portion is ultraviolet radiation. It is the Sun's emitted energy and Earth's position, not too close and not too far away, that make the planet livable.

TRAVELING ACROSS THE SKY?

Mythical gods like Apollo or Helios may have raced chariots or flown across the sky, but the Sun only appears to travel east to west each day because of our Earth-bound perspective and the planet's counterclockwise rotation. Each day as Earth rotates, the planet turns into and out of the Sun's light. At locations moving into the light, the Sun appears to rise in the east, and morning breaks. At locations moving out of the light, the Sun appears to set in the west, and it turns dark. Think of it as if Earth is moving as the Sun stays still (okay, the Sun isn't really staying still, but it helps to think of it that way).

SUMMER HEAT AND THE SUN?

Over the course of a year, 365 days, Earth orbits the Sun. The orbit is slightly elliptical, so that at certain times of the year Earth is a bit closer to the Sun than at others. *Is that why it's hotter in the summer?* That's a big, hot-star *no*. Because the Sun is so far away from Earth, slight changes in distance have little impact on Earth's weather. In fact, during summer in the Northern Hemisphere, Earth is at its most distant point from the Sun. It's all about Earth's tilt, the angle of the axis about which the planet spins. Earth's axis is about 23.5 degrees away from vertical, relative to the plane of its orbit around the Sun. So, as Earth orbits the Sun, part of Earth is tilted toward it and part is tilted away. In December, the Southern Hemisphere (summer) is tilted toward the Sun and the Northern Hemisphere (winter) away. In June, when Earth is on the opposite side of the Sun, the Northern Hemisphere is tilted toward the Sun (summer), and the Southern Hemisphere is tilted away (winter). And voilà—Earth's seasons.

WHAT ARE THOSE STRANGE RINGS AND RAINBOWS IN THE SKY?

A ring around the Sun (or Moon) is often thought a harbinger of snow or rain—or a forewarning of doom. In contrast, rainbows encircling the sun or perched in the sky offer hope or a good omen. Whatever you believe, there's also a scientific explanation for these sky-high phenomena.

The electromagnetic radiation or light emitted by the Sun has varying wavelengths. Visible light, what the human eye can see, has wavelengths between 0.000015 and 0.00003 inches (380 to 700 nanometers). This range includes colors with increasing wavelengths going from purple and blue to green, yellow, orange, and red. Typically they're all seen together as white light. But if visible light is refracted or reflected, it can be split into its various color components, creating a rainbow effect.

Halos or rings about the Sun (or Moon) occur when light is refracted (bent) through ice crystals that make up high, thin cirrus clouds, which often appear wispy.

Similarly, small rainbows on the right, left, or both sides of the Sun, known as sun dogs, are surprisingly common but often go unseen (plate 18). They too are due to light refracting through ice crystals in cirrus clouds.

Sometimes a shaft of sunlight rises above the Sun at sunset or sunrise. It's not a portal to another world but a sun pillar, the reflection of light off slowly falling ice crystals within cirrus clouds or very cold air.

The more you look up, the more likely you are to see some of these super-cool spectacles in the sky. It's a free show that's simply amazing!

IS THE GREEN FLASH REAL OR IMAGINED?

The green flash is a short burst of green light seen just above the horizon at sunset. *Is it real, imagined, or perhaps the result of a few too many green flash cocktails?* Myths surrounding the green flash suggest it might have something to do with a soul departing or matters of the heart. Though rare to see, the green flash is a real phenomenon and explainable by . . . science. It's a little complicated, so here's a simplified version.

A momentary flash of green at sunset can occasionally be seen if conditions are just right—typically, a clear sky with the ocean on the horizon. Two phenomena are at play: the dispersion of sunlight and a mirage. As the Sun slips below the horizon, Earth's atmosphere refracts and disperses the sunlight like a prism. Long-waved light, such as red and orange, is bent too little to be seen over the horizon; short-waved blue and violet light is bent enough but quickly scattered. Green light is just right, the Goldilocks of wavelengths, to be ever-so-briefly observed. It takes a mirage created by the open ocean on the horizon to magnify the differences in the refracted wavelengths so the phenomenon can be seen.

Spend enough time at sea, especially in the tropics, and you're more likely to observe a green flash (Ellen). But there's another way to see it. Years ago, a speaker (our sincere apologies as one of us has forgotten his name) gave a stunning talk about the green flash at the Woods Hole Oceanographic Institution. He had set up a time-lapse camera on the balcony of his Florida condo to capture green flashes at sunset and sunrise (they occur at both). In this case, the old saying "a picture is worth a thousand words" could not be truer. Mesmerized by the speaker's videos, many in the audience surely left as green flash evangelists. If in doubt, there are numerous credible images of the green flash available on the web.

ARE SOLAR ECLIPSES DANGEROUS?

With proper eye protection, solar eclipses are not dangerous. Experts assure us you will not die, go blind, or, as someone once suggested, change sex if you observe one safely.

Throughout the ages, solar eclipses have captivated people's attention and inspired cultural mythologies. They've been cited as a forewarning of sickness, doom, or tragedy. Explanations ranged from the wrath of God to the work of a monster, a giant frog, or a wolf that ate the Sun. Solar eclipses have also been characterized as the result of an intimate relationship between the Moon and Sun. Though the science behind solar eclipses may be less creative, it is no less fascinating.

A solar eclipse occurs when the Moon, in passing between Earth and the Sun, blocks the Sun's light. *How can the Moon, which is four hundred times smaller than the Sun, block it out?* The Moon is also about four hundred times closer to Earth than is the Sun. Part of the magic of an eclipse is that it takes an exact orbital alignment, combined with the special ratios of size and distance of the Moon and Sun, to make it happen.

When the Moon blocks 100 percent of the Sun, it is a total solar eclipse. A shadow is cast on Earth, and as the planet rotates (and the Moon orbits), it creates a trail called the path of totality. It is only within the path of totality that a person can experience full darkness during a total solar eclipse. There, for just a few minutes, day turns to night. Animals may momentarily behave as if night has fallen, and a noticeable change in temperature may occur. Dave has experienced totality and attests that it's mesmerizing—some would say, a spiritual experience. It is also the one time during an eclipse when protective glasses may be removed to safely view totality and see the full spectacle.

Total solar eclipses also offer the rare opportunity to see the Sun's outer atmosphere, or corona. Usually, the corona isn't visible because of the overpowering brilliance of the Sun. But during a total eclipse, at just the right moment, the corona appears like a dazzling glow surrounding the blacked-out Sun, sometimes with slivers of light streaming or arching outward (plate 19).

When the Moon blocks out less than 100 percent of the Sun, it is a partial solar eclipse. One type of partial eclipse is

THE SUN

an annular solar eclipse. This happens when the Moon passes between the Sun and Earth but is at or near its farthest point from Earth (apogee). Because of the greater distance, the Moon appears smaller than the Sun and doesn't block it completely.

Knowing the orbits of the Moon and Earth relative to the Sun, scientists can calculate when partial, annular, and total solar eclipses will occur and where best to see them. On April 8, 2024, a total solar eclipse occurred in the United States. Millions of people traveled to the path of totality, praying for a clear sky to experience full darkness and to catch a glimpse of the

Wear official sun protection glasses to look directly at the Sun!

Sun's luminous corona. In advance, transportation officials and cell phone providers prepared for heavy traffic on the roads and on the airwaves, hoping neither got jammed up. For those who witnessed it, it may have been both life-changing and a once-in-a-lifetime event, because the next total solar eclipse visible in the United States won't be until August 23, 2044. But look for an annular solar eclipse on October 2, 2024, and partial solar eclipses on March 29 and September 21, 2025.

Another reminder: To observe the Sun safely during an eclipse, proper safety gear is a must. NASA or other approved eclipse glasses with the number ISO 12312-2 should always be worn, or an indirect method of observation used.

SPACE WEATHER: WHAT IS IT AND WHY SHOULD WE CARE?

Space weather is not the weather in a galaxy far, far away or beyond the Sun; it refers to changing conditions in the area between the Sun and Earth. Like Earth's weather, space weather can be quiet and calm or active and stormy. It can also impact or disrupt activities on Earth and in space.

Space weather is driven by the Sun, a super-hot, super-big spinning ball of electrically charged gas or plasma, which is created by the extreme heat of the Sun and the collisions of fast-moving gas particles that knock electrons off atoms. The roiling plasma and spinning Sun produce enormous, chaotic magnetic fields that get twisted and tangled. This generates huge amounts of energy, which creates solar wind. Solar wind streams out from the Sun's corona and, if directed toward

Earth, strikes the planet's magnetic field. Most of the time, Earth's magnetic field acts like a shield against the solar wind. But not always.

Increased activity on the surface of the Sun produces sunspots (dark, cool areas where magnetic field lines are concentrated) and can trigger solar storms in the form of solar flares and coronal mass ejections (CMEs). Solar flares are eruptions of light and charged particles generated by the release of magnetic energy; whereas coronal mass ejections are large clouds of solar plasma ejected into space after a solar eruption. Both can cause the solar wind to intensify, flow along the lines of Earth's magnetic field, and pass into the planet's atmosphere at the poles. When increased solar activity, such as flares or CMEs, disturbs or disrupts Earth's magnetic field, it's called a geomagnetic storm. Strong geomagnetic storms can disrupt global communications and navigation systems and cause electrical blackouts. Events that increase the radiation output from the Sun are especially dangerous to astronauts.

Seen high in the night sky, colorful auroras are also a result of intense solar activity. These glowing waves of green, red, or even purple light occur principally above the northern and southern poles and are thus called the northern and southern lights, respectively (plate 20).

Auroras occur when solar winds collide with Earth's magnetic field and charged particles accelerate explosively into Earth's atmosphere. Think of it as energized particles surfing the lines of the magnetic field toward the poles. Once in the upper atmosphere, the charged particles (principally electrons) smash into oxygen and nitrogen atoms and molecules. The atoms get a boost in energy. When they relax back to their previous state, this energy is released as a burst of light. As NASA describes it, "When we see the glowing aurora, we are watching a billion individual collisions, lighting up the magnetic field lines of Earth." With increased solar activity and stronger CMEs, auroras may be pushed equatorward and seen at lower latitudes.

While observing sunspots on September 1, 1859, British astronomer Richard Carrington saw a bright flash of light. One day later, a strong geomagnetic storm struck Earth, disrupting and damaging telegraph networks across the globe. Telegraph operators received startling shocks, and lines caught on fire. Ship compasses went awry, and stunning auroras were seen as far south as the Caribbean and Mexico, as well as in Australia. The result of a coronal mass ejection, the incident is now known as the Carrington event. We asked NASA scientists what would happen if the Carrington event happened today.

Astrophysicist Dr. Michael Kirk acknowledged that it's a bit difficult to say exactly, but in the worst-case scenario with a direct impact on North America, there would probably be significant impacts, including rolling blackouts on the power grid and damage to transformers, and many small satellites would be "toast." For several days, GPS signals might be useless, meaning we'd have to resort to old-fashioned methods for navigation. Air traffic worldwide could be grounded for at least a day, possibly more. But Dr. Kirk suggests the good news is that with our space weather infrastructure and forecasting capabilities, there should be more warning than in the past and preparations can be made, such as powering down and reorienting large satellites.

Dr. Lika Guhathakurta adds that if such a strong geomagnetic storm hit Earth today, in-space astronaut operations would also be affected. But she notes that we've already had and learned from several events probably as strong or nearly as large as the Carrington event. In 1989, a strong geomagnetic storm took out power grids across the United States, and in Quebec, Canada, millions were left in the dark for hours. In 2022, intensified solar activity and the resulting geomagnetic storm laid waste to thirty-eight Starlink satellites.

NASA scientists are working hard to better understand heliophysics, including auroras and space weather, while at NOAA, experts produce an official government forecast on space weather and issue warnings. Alerts are particularly important for operators of critical infrastructure, such as power companies, transportation officials, and satellites. As we head into a solar max, it is an especially busy and important time for all involved.

Like Earth, the Sun has a north and a south magnetic pole, but every eleven years or so, they reverse, and the Sun cycles from quiet and calm (solar minimum) to violently active (solar maximum) (plate 21). During solar minimums, when the Sun's magnetic field weakens, astronauts face an increased risk of high-dose radiation from cosmic rays. At the solar max, sunspots, solar flares, and CMEs increase, and space weather events escalate. During the Sun's solar cycles, its total radiation output varies only slightly, about 0.15 percent (not enough to have a long-term impact on Earth's climate).

THE SUN

The last solar minimum was December 2019, and the next solar max is expected in late 2024 or sometime in 2025. Experts are closely monitoring the Sun's activity, issuing space weather forecasts, and sending out alerts if radiation or geomagnetic storms threaten space-based systems, astronaut activities, or on-the-ground operations.

OBSERVING THE SUN AS NEVER BEFORE

With multiple missions focused on the Sun, NASA and its partners are obtaining unprecedented data and never-before-seen imagery that is revolutionizing our understanding of the Sun, the Sun-Earth connection, and space weather. The STEREO (Solar TErrestrial RElations Observatory) mission with its two nearly identical spacecraft, one put into orbit about the Sun slightly ahead of Earth (STEREO-A) and one behind (STEREO-B), was launched in 2006. In 2010, the Solar Dynamics Observatory (SDO) blasted off; it was followed by the Parker Solar Probe in 2018 and the Solar Orbiter in 2020. Outfitted with high-tech instruments, sensors, and cameras, these spacecraft, and more to come, are getting closer to the Sun than ever before, collecting data in places never before sampled, and providing views from all new perspectives. Scientists are using the data to study a wide range of heliophysics topics, including the Sun's magnetic field, surface, and corona, along with solar winds, sunspots, coronal mass ejections, and strange features such as puzzling ropey arcs of plasma (coronal loops) and the mysterious solar hedgehog, an area more than 15,000 miles across discovered in the Sun's atmosphere with radiating spikes of hot and cold gas. A few highlights showcase the awesome results so far.

In February 2011, STEREO-A and STEREO-B were at a 180° angle to each other. For the first time, the front and back sides of the Sun were viewed and imaged simultaneously, and humanity saw the Sun in its true form—a sphere. Unfortunately, STEREO-B was lost in 2014, but STEREO-A continues its journey in orbit about the Sun, collecting imagery and data and working in concert with other spacecraft.

In February 2021, the Solar Orbiter recorded its first solar eruption, providing a grainy black-and-white video of solar wind, dust, and cosmic rays, with a CME thrown in for added excitement. In December 2021, the Parker Solar Probe became the first spacecraft to enter and sample the Sun's corona, which has long baffled scientists because, mysteriously, it is hotter than the Sun's surface. On another pass through, solar activity was surprisingly quiet and the escape of charged particles slow. Excited scientists described it as like going through the eye of a storm. In June 2023, the Parker team reported data suggesting that holes in the Sun's corona may be where super-fast solar winds originate. In December 2024, the Parker Solar Probe is expected to make its closest approach to the Sun, orbiting at 430,000 miles per hour and getting as near as 3.8 million miles to the surface. So, stay tuned.

As spacecraft sample and view the Sun as never before and technology improves, our understanding about our world and the star that powers it is advancing rapidly. Perhaps soon, your local weathercast will include images of the Sun and forecasts for when powerful solar storms are headed our way.

THE SUN

IS THE SUN THE DRIVING FORCE BEHIND RECENT CLIMATE CHANGE?

For more than forty years, scientists have used satellite technology to monitor the Sun's energy output. Over that period, slight variations have occurred in association with the eleven-year solar cycle. But averaged over time, not only has incoming total solar radiation not increased, it appears to be trending slightly downward. The takeaway: Earth's recent global temperature rise is not driven by increased solar output. For argument's sake, if it were, Earth's upper atmosphere, which is most directly affected by solar radiation, would be getting increasingly hot. Measurements show that the upper atmosphere is instead getting cooler. The lower atmosphere, on the other hand, is heating up, as expected as a result of increasing carbon dioxide levels due to anthropogenic (human) activities.

■ ■ ■

Like other stars, the Sun shines with its own light. It illuminates the world around us, including other planets and asteroids. It supports and fuels life on Earth, fascinates and bedazzles us. Please see the back of the book for our sources. We hope you'll want to learn more about our world's only daytime star, the Sun, along with heliophysics, eclipses, space weather, and more.

Chapter Twelve

INFORMATION MIXOLOGY

"**HEY, ELLEN.** I heard you used to scuba dive in Government Cut to lie on the bottom and watch cruise ships go into the Port of Miami."

"Whaat?"

After winning a grant from NASA while working at NBC4 in Washington, DC, Dave was asked as a "space expert" to take the lead on a news story about the reentry of a satellite with a nuclear battery. Whaat?

These stories are examples of how misinformation and misunderstandings can be born, evolve, and spread. Ellen entered graduate school with extensive field and diving experience. Mix that with studying at renowned coral reef geologist Dr. Robert Ginsburg's laboratory on Fisher Island adjacent to Government Cut in Miami, Florida (where vessels enter the port), and it turned into the diving-to-watch-cruise-ships story. With Dave's grant from NASA and a weather set with a NASA logo on it, he became an instant space expert. It's information mixology—stirring together a few facts to get a new and entertaining concoction. It may be easy to take in and seem reasonable but often comes with a chaser of misinformation or misunderstanding.

Perhaps the story about diving in Government Cut was simply a joke that, as it passed from one person to the next, evolved into a "true story." Or maybe the story got embellished with each telling. And if the source was a trusted friend or colleague, there may have been no reason to doubt its veracity.

Because Dave had a project funded by NASA, people in the newsroom somehow assumed he knew all about space. But he's not an astrophysicist, an astronaut, or a space engineer. An association with an expert or an organization does not magically bestow years of specialized knowledge and experience.

Get expert advice from the right expert.

And expertise in one area does not mean expert proficiency in another. We wouldn't ask an electrician to replumb the toilet. Talk about a crapshoot. Nor would we ask a veterinarian for advice about climate change.

Dr. Barb Kirkpatrick, a senior adviser at Texas A&M and the Gulf of Mexico Ocean Observing System, attended an open house at Mote Marine Laboratory in Sarasota, Florida. She relayed that upon learning that phytoplankton use sunlight to photosynthesize and produce about half the oxygen in our atmosphere, someone exclaimed, "Oh, so that's why I feel sluggish on cloudy days, there's less oxygen in the air." Whaat?

That's a thumbs-up for understanding that oxygen production by phytoplankton is important for life support on the planet; thumbs-down for misinterpreting cause and effect. To be clear, a decrease in oxygen production by phytoplankton on a cloudy day will not affect your breathing or energy level. If daily changes in photosynthesis dramatically affected atmospheric oxygen levels, at night we'd all die. There's enough oxygen in the atmosphere so that day-to-day changes due to photosynthesis are inconsequential to us, though we can measure them.

How information or data are interpreted can lead to misunderstandings, often as they pertain to cause and effect. Phytoplankton produce oxygen through photosynthesis, which uses sunlight. When it's cloudy, there's less sunlight (cause), so . . . there must be less oxygen in the atmosphere, which will affect our breathing (effect). The cause was correct, but the interpretation of the effect was not.

We like apples, and medical experts say eating apples is healthy. But is eating apple pie every day also healthy? Not so much. Here's another one. Plants use carbon dioxide to grow, so more carbon dioxide in the atmosphere is a good thing. Not so fast. As discussed in the chapter on climate change, plant growth is also controlled by other factors such as temperature, water availability, and soil conditions. Increased carbon dioxide in the atmosphere is causing extreme heat, drought, wildfires, and flooding, and then there's the bad that comes with super-potent poison ivy and, for allergy sufferers, more pollen.

Misleading headlines and social media posts readily give birth to misinformation and cause the spread of misunderstandings as well as major headaches for scientists. Take Hurricane Irma in 2017. Here are a few of the headlines and posts that sent the National Hurricane Center and many meteorologists into firefighting mode: *"Category 6? If Hurricane Irma Becomes the Strongest Hurricane in History, It Could Wipe Entire Cities Off the Map"* and *"Hurricane Irma Could Be a Category 6 by the Time It Hits the East Coast."* There is no Category 6 on the Saffer-Simpson Hurricane Wind Scale! Posts on Facebook also suggested the "long-range" forecast track indicated Irma was headed for Texas. At the time, forecasts did not extend greater than five days out. There was no credible "long-range" prediction, nor was the storm forecast to aim for Texas. Hurricane Irma made its first U.S. landfall in the Florida Keys as a Cat 4 and then moved north as a Cat 3. It was a powerful and destructive hurricane, but not a Cat 6 (again, no such thing). At one point there was uncertainty in long-range modeling as to when the storm would turn north (in the Gulf of Mexico), but this was in the modeling, not the official NHC five-day forecast, which, as the atmospheric steering patterns became clearer, took Irma into Florida.

INFORMATION MIXOLOGY

Buzzwords aren't always the right choice.

Repetition is another means by which misinformation can be born and spread. Sometimes a phrase or piece of information that is founded in solid science is used so often and in so many ways, the true meaning gets lost in translation. If heard often enough, even things with no basis in fact can become believable. This is particularly true if the information confirms a person's preexisting ideas or beliefs, a phenomenon called confirmation bias. Hurricanes again offer a few good examples. People want to believe that a hurricane won't hit where they live, so they look for something to confirm that idea . . . like Indian burial mounds somehow diverting storms. A television viewer sees the forecast of a hurricane five days out, and where their family lives is not in the cone of uncertainty. Rather than watch for updates, the person fixes on that forecast because it is the one they want.

Confirmation bias is capitalized on and promoted by organizations, corporations, and especially advertisers. By tracking views and likes on the internet, along with your personal information, social media and technology companies can use algorithms to push topics, products, or ideas that fit individual biases. And this can lead down the proverbial rabbit hole of misinformation or, when it is purposely done and misleading, disinformation.

Cherry-picking is another effective means of creating and spreading misinformation or disinformation. We are huge *Jurassic Park* fans. So, when Michael Crichton's book *State of Fear* came out, booklover Ellen was especially excited. However, she was soon wishing a *T. rex* had chomped, chewed, and swallowed every single copy. Throughout the book, Crichton used tidbits of real scientific data or subsets of larger datasets to make a point. This is cherry-picking. For instance, he suggested that temperature records from selected stations that show cooling are representative of global temperature monitoring—not. He used snippets of quotes from leading scientists out of context and misrepresented data on sea-level rise, the urban heat island effect, the impacts of land use, and more. But it's a novel of fiction, so what's the big deal? Except there is a message at the end meant to convince readers that global warming is not a real problem (just like in the book). We still love *Jurassic Park* but were sad and infuriated that Crichton used his celebrity author status to spread disinformation. Then again, as noted before, we wouldn't hire an electrician to fix the toilet, nor look to a fiction author to provide a credible analysis of climate data.

When discussing how misleading information or disinformation is spread, it is worth repeating what author Larry Kusche identified as helping to manufacture the mystery surrounding the Bermuda Triangle: people who (1) avoid fact checking, (2) add some imaginative speculating, (3) include a few unanswered questions, and (4) use a lot of exclamation points!

Creating a false balance of opinions can also confuse people or help spread scientific disinformation. One way to do this is to give equal weight to two sides of a story when, in reality, one side represents only a small (though vocal) minority. Part of the issue here is where the information comes from. And this brings us to the root source of many science-based misunderstandings: What is the source of the information involved? That's the focus of the next chapter.

■ ■ ■

We've offered a few examples and dipped our toes into the topic of the origins of misinformation and misunderstandings and how they spread. We provide our sources at the back of the book and encourage you to explore this topic in greater detail.

Chapter Thirteen

SHOW US THE DATA

DINOSAURS AND HUMANS never coexisted on Earth. *Never.* Dinosaurs went extinct more than sixty million years before humans inhabited the planet. But if you want to argue with us that they lived at the same time, *show us the data.* Show us the scientific evidence to prove your point. Do you have fossils or bones of humans and dinosaurs in the same geologic formation? Nope. How about a dinosaur fossil with evidence it fed on humans? Uh-uh. Direct observations or video maybe? Nada. Cave paintings, myths, drawings, or written stories don't count as scientific evidence. Nor do the Flintstones or posts on TikTok, Instagram, and Facebook. We might then ask: Where did you learn that dinosaurs and humans coexisted? What was your source?

As scientists, for topics founded in science, we base our opinions on data: collected and analyzed factual information, observations, and repeatable experimental results. We look to credible experts in specific fields for their science-based opinions—not their beliefs or ideology, but their opinions as scientists based on evidence and data. We are open to other opinions and happy to listen. But often our first question is: Where did your information come from? What's your source? This is followed by: Show us the data.

Today, people have access to more information than ever before—so much so that it can be overwhelming, like trying to take a sip from a firehose. The glut of data and information can be mentally exhausting and leave a person struggling to figure out whom to trust and how to find truth and clarity.

According to recent surveys, many Americans report getting their scientific information from general news sources. In cost-cutting efforts, most national and local news outlets have done away with their science-focused positions, resulting in less science coverage and less accuracy. Often journalists covering science-based stories have little to no background in science and little knowledge of how science works. Headlines are frequently the responsibility of someone who didn't even research, write, or produce the story involved and are often created to grab attention rather than to accurately reflect content. For example, in September 2023, when Massachusetts was hit with heavy rain from a nearby low-pressure area and frontal system, a magazine posted the headline "Hurricane Lee Is Pounding New England So Hard That Hundreds Are Being Evacuated and a 15-Foot Dam Is in Danger of Collapse." Hurricane Lee was 1,200 miles away! The magazine article on the flooding was accurate, but the headline was either hastily done or contrived to attract more eyeballs.

The good news is that surveys also suggest the public still has trust in science and scientists and see specialty sources (science-focused) as likely to be more accurate. It's sometimes hard not to be influenced by highly visible and vocal reports suggesting a public lack of trust in science or experts, but larger-scale polling indicates this is not a majority viewpoint. Surveys also suggest that people who participate in science activities have more trust in and understanding of science.

While people joke about the accuracy of weather forecasts, weather is still the number one reason people watch local news. Local and national meteorologists are beloved, and research shows that in many cities across the nation, they are often the most trusted personalities. More and more, they are working with experts to provide science information beyond just the weather.

With so many sources, means of getting information, and even information about information, where should you go to find trusted science information? We love our friends and family, but they're not always reliable sources of information, especially science information. We enjoy watching the national news and following breaking headlines, but the science presented is often inaccurate or incomplete. Social media provide another way to stay up to date on news headlines or things happening around the world, but here misinformation can run wild, and some people are paid to foster confusion and promote disinformation. Influencers on TikTok are rarely experts in science. Scientific journals, magazines, and websites that publish the results of peer-reviewed research are good sources, but these are sometimes difficult to access and follow and tend to be highly technical unless you're an expert in the topic involved. So . . . what to do?

Number one, ask questions! Where does information come from? What is it based on? Is the source actual data or simply an idea, someone's belief, or just a theory? Is it a research study that has been published in a peer-reviewed science magazine or an off-the-cuff Op-Ed, blog, or TikTok post? Is an organization presenting the information, and if so, who are they and what do they do? Can I find other reliable sources that confirm the information?

Even for organizations we trust to provide good science information in easy-to-understand language, we look for the original story source, such as a link, a mention of the scientist involved, or the research study the story is based on. Wikipedia alone is not a good scientific source, but it can be used as a guide to find credible scientific sources (check the references). Members of the Earth Science Information Partnership (ESIP) are working on ways to identify data as trusted for use in rapid response operations. Think of it as a "trust label" for data.

Sometimes we encounter people who question the credibility of leading well-respected scientists or institutions. People have even suggested that scientists get rich by coming up with the results funding organizations want. Having been involved with the grant and publication process for years, we can assure readers that in credible science organizations, including the National Science Foundation, NASA, NOAA, USGS, and others, that's not how it works. There are usually careful checks on conflicts of interest and qualifications, and we can honestly report we know of no scientist who has gotten rich through federal grants.

Here are a few of the organizations we have worked with, rely on, and recommend for scientific information. It doesn't mean you shouldn't ask questions about their information or look for original data and multiple sources to confirm, or that there aren't other good sources, but these are a good place to start.

> American Association for the Advancement of Science (AAAS)
> American Geophysical Union (AGU)
> American Meteorological Society (AMS)
> Climate Central
> George Mason University: Center for Climate Communication
> Intergovernmental Panel on Climate Change (IPCC)
> National Academy of Sciences (NAS)
> National Aeronautics and Space Administration (NASA)
> National Oceanic and Atmospheric Administration (NOAA)
> Smithsonian Institution
> U.S. Geological Survey (USGS)
> U.S. Forest Service
> University Corporation for Atmospheric Research (UCAR)
> Yale School of the Environment

■ ■ ■

Additional sources are provided at the back of the book. We have not mentioned individuals here, but you can find many of our trusted personal sources in the references and in our acknowledgments.

Chapter Fourteen

QUESTIONS ANYONE? ANYONE?

WE ARE ESPECIALLY CURIOUS and tend to ask a lot of questions. At a recent American Meteorological Society meeting for broadcast meteorologists, Dave asked so many questions following talks he was, in dramatic fashion, given his own microphone. It was great for a hearty laugh by all (especially us). But later in the meeting, a young meteorologist stood up and nervously asked a question. She began by saying that Dave had inspired her and given her the courage to come up to the front of the room to ask her question. Asking questions is important.

We like to joke about some of the questions we and our colleagues are asked, but they're all good. Sometimes they make us laugh; other times they get us thinking about someone else's perspective or way of thinking. We honestly believe questions are the lifeblood of learning. They are critical to increasing knowledge, finding facts, weeding out mistakes, and learning how to communicate effectively.

We'll wrap this up by imploring everyone to ask more questions. When people give talks and no one asks a question, speakers can feel like no one was listening or engaged, even if the entire audience was thoroughly engrossed. And if you're thinking your question isn't smart enough—stop. Others

QUESTIONS ANYONE? ANYONE?

in the audience may be wondering the very same thing. Simple, outside-the-box questions can challenge us in science. They make us see our work through other eyes and to explain complicated things in easy-to-understand language. So go ahead and ask that question. Don't be shy or insecure. Your questions matter.

One last note: We hope that with this book, we've answered some of your questions about the ocean and the atmosphere. We also hope we've shown that the ocean and the atmosphere are not separate, battling entities at odds with one another but are intertwined or coupled on this wonderful and fascinating planet we call Earth.

The ocean and atmosphere are interconnected on planet Earth.

ACKNOWLEDGMENTS

FROM THE SEED of an idea to its release, working on and publishing a book is exciting, challenging, frustrating, exhilarating, exhausting, and above all else, it takes a team. To all of you who supported and worked with us on this book, we are enormously grateful.

We must begin with our friends, families, and colleagues who didn't shake their head or stare at us like we're numbskulls whenever we bent their ears about the book. Instead, they provided endearing and passionate encouragement. Thank you. Huge appreciation goes to Miranda Martin at Columbia University Press for supporting and jockeying the project through the publishing process, especially given its nontraditional tone for an academic press. Thanks to the board, production editor Kathryn Jorge, catalog manager Zachary Friedman, publicist Robyn Massey, wonderful cover designer Milenda Nan Ok Lee, and all the others at Columbia University Press for their

ACKNOWLEDGMENTS

assistance and support. We gratefully recognize the excellent copy editing by Peggy Tropp and editorial manager Ben Kolstad at KnowledgeWorks Global. And our heartful thank-you to Alece Birnbach for her amazing illustrations that took our ideas and made them oh-so-much better.

Writing popular science books is undoubtedly a labor of love, and rarely is funding available to support the task. We want to give special thanks to the Alfred P. Sloan Foundation and NASA Heliophysics for their financial support of the research and writing of the book.

Next up is our amazing network of friends and colleagues. Our sincerest gratitude for your expert advice, humorous stories, favorite and most frequently asked questions, and enthusiasm. Your input was invaluable and inspirational. Thank you. The list is a little long, but here goes. Thank you to NOAA administrator Dr. Rick Spinrad, NOAA National Weather Service director Ken Graham, the NOAA Ocean Exploration and Research Program and especially Emily Crum, NASA heliophysics experts Dr. Lika Guhathakurta, Dr. Michael Kirk, and Christina Milotte, Captain Peg Brandon, Dr. Jim White, Mike Seidel, Mike Bettes, John Morales, Dr. Edie Widder, Dr. Susan Humphries, Dr. Shirley Pomponi, NWS meteorologists Tanja Fransen and Doug Hilderbrand, Dr. Ken Lohman, Dr. Roger Hanlon, Dr. Barb Kirkpatrick, Bill Precht, Dr. Bob Hueter, Eugene Shinn, Dr. Brian McNoldy, and two men in a bar in Maine. We'd also like to acknowledge all those who helped make some of our favorite movies that are mentioned in the book (even if the science isn't all technically accurate): *Twister*, *Jurassic Park*, *The Mummy*, *Jaws*, and *The Day After Tomorrow*.

ACKNOWLEDGMENTS

Images often bring words on a page to life. Thank you to underwater photographer extraordinaire Stephen Frink, Dr. Tom Pluckhahn, NOAA's Ocean Exploration Program, NASA, meteorologist Jeff Berardelli, Dr. Edie Widder, Dr. Ken Lohman, Dr. Live Williamson, Alexandra Wen, Walter Lyons, and Frankie Lucena.

Thank you to the peer reviewers, meteorologist Paul Gross, geologist Dr. Tim Dixon, and someone who preferred to be anonymous, for your time and thoughtful comments.

Thank you also to our colleagues whose friendship and words cheered us on, including meteorologists Jim Cantore, Craig Setzer, Jay Trobec, Stephanie Abrams, and Jim Gandy, Bernadette Woods Placky, Dr. Steven Miller, Dr. David Shiffman, Debbi Stone and Virginia Wright-Placeres, and Sarah Falkowski.

A special thank you to my daughters Lindsay Jones and Heather Looney (Jones) for your support and love. See, I got your names in my first book, Dad!

If we've missed anyone, our apologies. We look forward to working with all of you and so many others in the years to come.

SOURCES AND ADDITIONAL INFORMATION

1. THE DEEP VAST SEA

Castello-Branco, Cristiana, Allen G. Collins, and Eduardo Hajdu. "A Collection of Hexactinellids (Porifera) from the Deep South Atlantic and North Pacific: New Genus, New Species and New Records." *PeerJ* 8 (2020): e9431. https://doi.org/10.7717/peerj.9431.

Chapell, Bill. "A U.S. Submarine Struck an Underwater Mountain Last Month, the Navy Says." National Public Radio, November 2, 2021. https://www.npr.org/2021/11/02/1051422572/navy-submarine-nuclear-collision-south-china-sea.

Cryptid. "Remembering the Fake Megalodon Documentary in Shark Week 2013." *ReelRundown*, March 10, 2023. https://reelrundown.com/tv/The-Shark-Week-2013-Fake-Megalodon-Documentary-Fiasco.

Davis, Josh. "Megalodon: The Truth About the Largest Shark That Ever Lived." Natural History Museum. https://www.nhm.ac.uk/discover/megalodon--the-truth-about-the-largest-shark-that-ever-lived.html.

Discovery. "Discovery Channel Celebrates 25 Years of Innovation, Excellence and Groundbreaking Nonfiction Programming." Press release, June 2, 2010. https://press.discovery.com/us/dsc/press-releases/2010/discovery-channel-celebrates-25-years-innovation-e/.

Ford, Michael, Nicholas Bezio, and Allen Collins. "*Duobrachium sparksae* (incertae sedis Ctenophora Tentaculata Cydippida): A New Genus and Species of

SOURCES AND ADDITIONAL INFORMATION

Benthopelagic Ctenophore Seen at 3,910 m Depth Off the Coast of Puerto Rico." *Plankton and Benthos Research* 15, no. 4 (2020): 296–305. https://www.jstage.jst.go.jp/article/pbr/15/4/15_P150401/_article/-char/en.

Heithaus, Michael. "Millions of Years Ago, the Megalodon Ruled the Oceans—Why Did It Disappear?" *The Conversation*, June 20, 2022. https://theconversation.com/millions-of-years-ago-the-megalodon-ruled-the-oceans-why-did-it-disappear-182841.

Lebreton, L., B. Slat, F. Ferrari, B. Sainte-Rose, J. Aitken, R. Marthouse, S. Hajbane, et al. "Evidence That the Great Pacific Garbage Patch Is Rapidly Accumulating Plastic." *Scientific Reports* 8, no. 4666 (March 2018). https://www.nature.com/articles/s41598-018-22939-w.

Moore, Charles. "Trashed: Across the Pacific Ocean, Plastics, Plastics, Everywhere." *Natural History*, November 2003. https://www.naturalhistorymag.com/htmlsite/master.html?https://www.naturalhistorymag.com/htmlsite/1103/1103_feature.html.

National Oceanic and Atmospheric Administration (NASA), Laboratory for Satellite Altimetry. "Altimetric Bathymetry." https://www.star.nesdis.noaa.gov/socd/lsa/AltBathy/.

National Oceanic and Atmospheric Administration (NOAA), Ocean Exploration. "How Does the Temperature of Ocean Water Vary?" https://oceanexplorer.noaa.gov/facts/temp-vary.html.

———. "A Magnificent New Sponge from the Deep Gets a Name." July 9, 2020. https://oceanexplorer.noaa.gov/news/oer-updates/2020/sponge-discovery.html.

———. "NOAA Scientists Virtually Discover New Species of Comb Jelly Near Puerto Rico." November 20, 2020. https://oceanexplorer.noaa.gov/news/oer-updates/2020/ctenophore.html.

———. "What Is a Seamount?" https://oceanexplorer.noaa.gov/facts/seamounts.html.

Nevala, Amy E. "Alvin Gets an Interior Re-design." *Oceanus*, August 6, 2010. https://www.whoi.edu/oceanus/feature/alvin-gets-an-interior-re-design/.

Shank, Tim, and Rick Chandler. "Alvin FAQs." Woods Hole Oceanographic Institution. https://www.whoi.edu/what-we-do/explore/underwater-vehicles/hov-alvin/faqs/.

Spector, Dina. "70 Percent of People Still Believe Megalodon Exists After Watching Discovery Channel's Fake Documentary." *Business Insider*, August 7, 2013. https://www.businessinsider.com/majority-of-discovery-channel-viewers-believe-megaldon-exists-2013-8.

SOURCES AND ADDITIONAL INFORMATION

Steingass, Sheanna. "Fishful Thinking: Five Reasons Why Mermaids Can't Physically Exist." *Deep Sea News*, October 13, 2013. https://www.deepseanews.com/2013/10/fishful-thinking-five-reasons-why-mermaids-cant-physically-exist/.

Vecchione, Michael. "*Deep Discoverer* Discovers a Very Deep, Ghostlike Octopod." National Oceanic and Atmospheric Administration (NOAA), Ocean Exploration, March 2, 2016. https://oceanexplorer.noaa.gov/okeanos/explorations/ex1603/logs/mar2/mar2.html.

Voosen, Paul. "'It's Just Mind Boggling.' More Than 19,000 Undersea Volcanoes Discovered." *Science*, April 19, 2023. https://www.science.org/content/article/it-s-just-mind-boggling-more-19-000-undersea-volcanoes-discovered.

Yancey, Paul. "Life Under Pressure—100 Elephants on Your Head." Schmidt Ocean Institute, November 12, 2014. https://schmidtocean.org/cruise-log-post/life-under-pressure-100-elephants-on-your-head/.

2. DANGEROUS MARINE LIFE

Bieler, Des. "A Pair of Kayakers in California Barely Avoid Being Swallowed by a Humpback Whale." *Washington Post*, November 4, 2020. https://www.washingtonpost.com/sports/2020/11/04/kayakers-humpback-whale-california/.

Bradford, Alina, and Patrick Pester. "Orcas: Facts About Killer Whales." *Live Science*, October 28, 2022. https://www.livescience.com/27431-orcas-killer-whales.html.

California State University, Long Beach. "Stingray Behavior and Biology." https://www.csulb.edu/shark-lab/stingray-behavior-and-biology.

Center for Coastal Studies. "Baleen Whales." https://coastalstudies.org/connect-learn/stellwagen-bank-national-marine-sanctuary/marine-mammals/cetaceans/baleen-whales/.

Florida Museum. "Discover Fishes: *Sphyraena barracuda*." https://www.floridamuseum.ufl.edu/discover-fish/species-profiles/sphyraena-barracuda/.

———. "International Shark Attack File." https://www.floridamuseum.ufl.edu/shark-attacks/.

Flounders, Lois. "Do Shark Repellents Work?" Save Our Seas Foundation. https://saveourseas.com/worldofsharks/do-shark-repellents-work.

Fox, Allen. "Venomous Sea Snakes That Charge Divers May Just Be Looking for Love." *Smithsonian*, August 19, 2021. https://www.smithsonianmag.com/science-nature/venomous-sea-snakes-charge-divers-may-just-be-looking-love-180978473/.

SOURCES AND ADDITIONAL INFORMATION

Fraser, Doug. "'I Was Completely Inside': Lobster Diver Swallowed by Humpback Whale Off Provincetown." *Cape Cod Times*, June 11, 2021. https://www.capecodtimes.com/story/news/2021/06/11/humpback-whale-catches-michael-packard-lobster-driver-mouth-proviencetown-cape-cod/7653838002/.

Helfman, Gene, and George H. Burgess. *Sharks: The Animal Answer Guide*. Baltimore: Johns Hopkins University Press, 2014.

Heupel, M. R., C. A. Simpfendorfer, and R. E. Hueter. "Running Before the Storm: Blacktip Sharks Respond to Falling Barometric Pressure Associated with Tropical Storm Gabrielle." *Journal of Fish Biology* 63, no. 5 (2003): 1357–63. https://doi.org/10.1046/j.1095-8649.2003.00250.x.

Hobson, Melissa. "Most Whales Can't Swallow a Human: Here's Why." *National Geographic*, June 15, 2021. https://www.nationalgeographic.com/animals/article/most-whales-cant-really-swallow-a-human-heres-why.

IMDb. "All Shark Movies." June 29, 2018. Accessed March 13, 2024. https://www.imdb.com/list/ls022606419/.

Lillywhite, Harvey B., Coleman M. Sheehy III, Harold Heatwole, Francois Brischoix, and David W. Steadman. "Why Are There No Sea Snakes in the Atlantic?" *BioScience* 68, no. 1 (2018): 15–24. https://doi.org/10.1093/biosci/bix132.

Morlock, Sarah. "16 Facts About Sea Snakes." PADI. https://blog.padi.com/sea-snake-facts/.

Naked Scientists. "Could a Human Survive Swallowing by a Whale?" June 27, 2010. https://www.thenakedscientists.com/articles/questions/could-human-survive-swallowing-whale.

National Oceanic and Atmospheric Administration (NOAA). "Sperm Whale." NOAA Fisheries. https://www.fisheries.noaa.gov/species/sperm-whale.

———. "What Is a Portuguese Man O' War?" National Ocean Service. https://oceanservice.noaa.gov/facts/portuguese-man-o-war.html.

Powell, Devin. "Shark Smell Myth Found Fishy." *Inside Science*, October 13, 2010. https://www.insidescience.org/news/shark-smell-myth-found-fishy.

Prager, Ellen. *Sex, Drugs, and Sea Slime: The Oceans' Oddest Creatures and Why They Matter*. Chicago: University of Chicago Press, 2011.

ReefQuest Centre for Shark Research. "Biology of Sharks and Rays." http://www.elasmo-research.org/education/topics/menu/topics_home.htm.

Shiffman, David. *Why Sharks Matter*. Baltimore: Johns Hopkins University Press, 2022.

Van Hoose, Natalie. "Why Are There No Sea Snakes in the Atlantic?" Florida Museum, November 29, 2017. https://www.floridamuseum.ufl.edu/science/why-are-there-no-sea-snakes-in-the-atlantic/.

Whale and Dolphin Conservation. "Facts About Orcas (Killer Whales)." https://us.whales.org/whales-dolphins/facts-about-orcas/.

3. JELLYFISH

Berwald, Juli. *Spineless: The Science of Jellyfish and the Art of Growing a Backbone.* New York: Riverhead, 2018.

MedlinePlus. "Jellyfish Stings." https://medlineplus.gov/ency/article/002845.htm.

Monterey Bay Aquarium. "Jellies." https://www.montereybayaquarium.org/animals/animals-a-to-z/jellies.

National Oceanic and Atmospheric Administration (NOAA). "What Are Jellyfish Made Of?" National Ocean Service. https://oceanservice.noaa.gov/facts/jellyfish.html.

Prager, Ellen. *Sex, Drugs, and Sea Slime: The Oceans' Oddest Creatures and Why They Matter.* Chicago: University of Chicago Press, 2011.

University of Hawaii. "Heat Trumps Cold in the Treatment of Jellyfish Stings." *News*, April, 14, 2016. https://www.hawaii.edu/news/2016/04/14/heat-trumps-cold-in-the-treatment-of-jellyfish-stings/.

WebMD. "Jellyfish Sting Treatment." https://www.webmd.com/first-aid/jellyfish-stings-treatment.

4. OTHER SEA CREATURES

Castro, Joseph. "How Do Dolphins Sleep?" *Live Science*, April, 14, 2014. https://www.livescience.com/44822-how-do-dolphins-sleep.html.

Guarino, Ben. "How Echolocation Lets Bats, Dolphins, and Even People Navigate by Sound." *Popular Science*, May 15, 2023. https://www.popsci.com/science/what-is-echolocation/.

Hecker, Bruce. "How Do Whales and Dolphins Sleep Without Drowning?" *Scientific American*, February 2, 1988. https://www.scientificamerican.com/article/how-do-whales-and-dolphin/.

SOURCES AND ADDITIONAL INFORMATION

Lohmann, Kenneth J., Catherine M. F. Lohmann, Llewellyn M. Ehrheart, Dean A. Bagley, and Timothy Swing. "Geomagnetic Map Used in Sea-Turtle Navigation." *Nature* 428 (2004): 909–10.

Lohmann, Kenneth J., Nathan F. Putman, and Catherine M. F. Lohmann. "The Magnetic Map of Hatchling Loggerhead Sea Turtles." *Current Opinion in Neurobiology* 22 (2012): 336–42. https://doi.org/10.1016/j.conb.2011.11.005.

Milius, Susan. "Octopuses Can "See" with Their Skin." *Science News*, May 20, 2015. https://www.sciencenews.org/article/octopuses-can-see-their-skin.

National Oceanic and Atmospheric Administration (NOAA). "Is It a Seal or Sea Lion?" NOAA Fisheries. https://www.fisheries.noaa.gov/feature-story/it-seal-or-sea-lion.

Prager, Ellen. *The Oceans*. New York: McGraw-Hill, 2000.

———. *Sex, Drugs, and Sea Slime: The Oceans' Oddest Creatures and Why They Matter*. Chicago: University of Chicago Press, 2011.

Ramirez, Desmond, and Todd H. Oakley. "Eye-Independent, Light-Activated Chromatophore Expansion (LACE) and Expression of Phototransduction Genes in the skin of *Octopus bimaculoides*." *Journal of Experimental Biology* 218, no. 10 (2015): 1513–20. https://doi.org/10.1242/jeb.110908.

Sargassum Information Hub. "What Is Sargassum." https://sargassumhub.org/about-sargassum/.

Sea Turtle Conservancy. "Information About Sea Turtles: General Behavior." https://conserveturtles.org/information-sea-turtles-general-behavior/.

Todd, Nicole. "Echolocation 101: How Dolphins See with Sound." Whale Scientists, October 9, 2020. https://whalescientists.com/echolocation-dolphins/.

University of North Carolina at Chapel Hill. The Lohmann Lab. https://lohmannlab.web.unc.edu.

Whale and Dolphin Conservation. "How Do Whales and Dolphins Sleep?" https://us.whales.org/whales-dolphins/how-do-whales-and-dolphins-sleep/.

5. CORAL REEFS

Atlantic and Gulf Rapid Reef Assessment. "Coral Disease Outbreak." https://www.agrra.org/coral-disease-outbreak/.

Corals of the World. "Reefs." http://www.coralsoftheworld.org/page/reefs/.

Cornwall, Warren. "After Mass Coral Die-off, Florida Scientists Rethink Plan to Save Ailing Reefs." *Science*, February 7, 2024. https://doi.org/10.1126/science.zpqtoki.

SOURCES AND ADDITIONAL INFORMATION

Costanza, Robert, Rudolf de Groot, Paul Sutton, Sander van der Ploeg, Sharolyn J. Anderson, Ida Kubiszewski, Stephen Farber, and R. Kerry Turner. "Changes in the Global Value of Ecosystem Services." *Global Environmental Change* 26 (2014): 152–58. https://doi.org/10.1016/j.gloenvcha.2014.04.002.

Ferrario, Filippo, Michael W. Beck, Curt D. Storlazzi, Fiorenza Micheli, Christine C. Shepard, and Laura Airoldi. "The Effectiveness of Coral Reefs for Coastal Hazard Risk Reduction and Adaption." *Nature Communications* 5 (2014): 3794. https://doi.org/10.1038/ncomms4794.

Ferse, Sebastian C. A., Margaux Y. Hein, and Lena Rölfer. "A Survey of Current Trends and Suggested Future Directions in Coral Transplantation for Reef Restoration." *PLoS ONE* 16, no. 5 (2021): e0249966. https://doi.org/10.1371/journal.pone.0249966.

Hein, M. Y., I. M. McLeod, E. C. Shaver, T. Vardi, S. Pioch, L. Boström-Einarsson, M. Ahmed, and G. Grimsditch. "Coral Reef Restoration as a Strategy to Improve Ecosystem Services: A Guide to Coral Restoration Methods." Nairobi, Kenya: United Nations Environment Program, 2020. https://www.icriforum.org/wp-content/uploads/2021/01/Hein-et-al.-2020_UNEP-report-1.pdf.

Henkel, Timothy P. "Coral Reefs." *Nature Education Knowledge* 3, no. 10 (2010): 12. https://www.nature.com/scitable/knowledge/library/coral-reefs-15786954/.

Hoegh-Guldberg, Ove, William Skirving, Sophie G. Dove, Blake L. Spady, Andrew Norrie, Erick F. Geiger, Gang Liu, Jacqueline L. De La Cour, and Derek P. Manzello. "Coral Reefs in Peril in a Record-Breaking Year." *Science* 382, no. 6676 (2023): 1238–40. https://www.science.org/doi/10.1126/science.adk4532.

Hubbard, Dennis K., and David Scatura. "Growth Rates of Seven Species of Scleractinean Corals from Cane Bay and Salt River, St. Croix, USVI." *Bulletin of Marine Science* 36, no. 2 (1985): 325–38. https://www.aoml.noaa.gov/general/lib/CREWS/Cleo/St.%20Croix/salt_river3.pdf.

Kench, Paul S., Edward P. Beetham, Tracey Turner, Kyle M. Morgan, Susan D. Owen, and Roger F. McLean. "Sustained Coral Reef Growth in the Critical Wave Dissipation Zone on a Maldivian Atoll." *Communications Earth & Environment* 3, no. 9 (2022). https://doi.org/10.1038/s43247-021-00338-w.

Montaggioni, Lucien F. "History of Indo-Pacific Coral Reef Systems Since the last Glaciation: Development Patterns and Controlling Factors." *Earth-Science Reviews* 71, no. 1–2 (2005): 1–75. https://doi.org/10.1016/j.earscirev.2005.01.002.

SOURCES AND ADDITIONAL INFORMATION

National Oceanic and Atmospheric Administration (NOAA). "Coral Cores: Ocean Timelines." Flower Garden Banks National Marine Sanctuary. https://flowergarden.noaa.gov/science/coralcores.html.

Prager, Ellen. *Sex, Drugs, and Sea Slime: The Oceans' Oddest Creatures and Why They Matter*. Chicago: University of Chicago Press, 2011.

Secore International. "Coral Reproduction." https://www.secore.org/site/corals/detail/coral-reproduction.15.html.

Souter, D., S. Planes, J. Wicquart, M. Logan, D. Obura, and F. Staub, eds. "Status of Coral Reefs of the World: 2020 Report." Global Coral Reef Monitoring Network (GCRMN) and International Coral Reef Initiative (ICRI), 2021. https://doi.org/10.59387/WOTJ9184.

Toth, Lauren T., Travis A. Courtney, Michael A. Colella, Selena A. Kupfner Johnson, and Robert R. Ruzicka. "The Past, Present, and Future of Coral Reef Growth in the Florida Keys." *Global Change Biology* 28, no. 17 (2022): 5294–5309. https://doi.org/10.1111/gcb.16295.

6. SUPERNATURAL, SUSPICIOUS, OR SCIENCE

Atmospheric Optics. "Inferior Mirage Green Flash." September 16, 2023. https://atoptics.co.uk/blog/inferior-mirage-green-flash-2/.

Bayartsogt, Khalium, and Emily Feng. "Buried Alive in Mongolia's Worst Sandstorms in a Decade." National Public Radio, May 30, 2021. https://www.npr.org/sections/goatsandsoda/2021/05/30/1000530563/buried-alive-in-mongolias-worst-sandstorms-in-a-decade.

Beal, L. M., and K. A. Donohue. "The Great Whirl: Observations of Its Seasonal Development and Interannual Variability." *Journal of Geophysical Research: Oceans* 118, no. 1 (January 2013): 1–13. https://doi.org/10.1029/2012JC008198.

Cooper, Helen. "Illuminating the Facts of Deep-Sea Bioluminescence." Monterey Bay Aquarium, November 15, 2020. https://www.montereybayaquarium.org/stories/bioluminescence.

EarthSky. "What's a Green Flash and How Can I See One?" December 7, 2023. https://earthsky.org/earth/can-i-see-a-green-flash/.

Federal Aviation Administration. https://www.faa.gov/sites/faa.gov/files/regulations_policies/policy_guidance/envir_policy/contrails.pdf.

Friend, Duane. "What Is a Microburst?" University of Illinois Urbana–Champaign, October 28, 2021. https://extension.illinois.edu/blogs/all-about-weather/2021-10-28-what-microburst.

Gabbert, Bill. "Researchers Try to Shed New Light on Weather Related to 19 Firefighter Deaths." Wildfire Today, March 15, 2022. https://wildfiretoday.com/2022/03/15/researchers-try-to-shed-new-light-on-weather-related-to-19-firefighter-deaths/.

Haddock, S. H. D., C. M. McDougall, and J. F. Case. "The Bioluminescence Web Page." https://biolum.eemb.ucsb.edu/.

Hurricanes: Science and Society. "Brief History of Hurricane Forecast Models." https://hurricanescience.org/science/forecast/models/modelshistory/index.html.

Jabr, Ferris. "Gleaning the Gleam: A Deep-Sea Webcam Sheds Light on Bioluminescent Ocean Life." *Scientific American*, August 5, 2010. https://www.scientificamerican.com/article/edith-widder-bioluminescence/.

———. "The Secret History of Bioluminescence." *Hakai Magazine*, May 10, 2016. https://hakaimagazine.com/features/secret-history-bioluminescence/.

Kusche, Larry. *The Bermuda Triangle Mystery Solved*. New York: Prometheus, 1986.

Lallanilla, Marc. "What Is Dew Point?" *Live Science*, February 11, 2014. https://www.livescience.com/43269-what-is-dew-point.html.

Latz Laboratory. "Bioluminescence Questions and Answers." Scripps Institution of Oceanography, University of San Diego. https://latzlab.ucsd.edu/bioluminescence/bioluminescence-questions-and-answers/.

Martini, Séverine, and Steven H. D. Haddock. "Quantification of Bioluminescence from the Surface to the Deep Sea Demonstrates Its Predominance as an Ecological Trait." *Scientific Reports* 7, no. 45750 (2017). https://doi.org/10.1038/srep45750.

Melzer, B. A., and T. G. Jensen. "Evolution of the Great Whirl Using an Altimetry-Based Eddy Tracking Algorithm." *Geophysical Research Letters* 46, no. 8 (April 2019): 4378–85. https://doi.org/10.1029/2018GL081781.

Mulvihill, Amy. "Dream Boat: The Tragic Last Voyage of the *Pride of Baltimore*." *Baltimore Magazine*, May 2016. https://www.baltimoremagazine.com/section/artsentertainment/the-last-tragic-voyage-of-the-pride-of-baltimore/.

Nagai, Takeyoshi. "The Kuroshio Current: Artery of Life." *Eos* 100, August 27, 2019. https://doi.org/10.1029/2019EO131369.

Nalewicki, Jennifer. "Puerto Rico's Bioluminescent Bays Are Brighter Than Ever." *Smithsonian*, April 6, 2022. https://www.smithsonianmag.com/travel/puerto-ricos-bioluminescent-bays-are-brighter-than-ever-180979874/.

NASA Scientific Visualization Studio. "The Science of Monsoons." July 7, 2016. https://svs.gsfc.nasa.gov/12303.

SOURCES AND ADDITIONAL INFORMATION

National Environmental Satellite, Data, and Information Service. "Fifty Years After Hurricane Camille, NOAA Satellites Keep U.S. Weather Ready." August 16, 2019. https://www.nesdis.noaa.gov/news/50-years-after-hurricane-camille-noaa-satellites-keep-us-weather-ready.

National Science Foundation. "Discovery of Microbursts Leads to Safer Air Travel." June 25, 2003. https://beta.nsf.gov/news/discovery-microbursts-leads-safer-air-travel.

National Weather Service. "Altocumulus Standing Lenticular Clouds." https://www.weather.gov/abq/features_acsl.

———. "Dew Point vs. Humidity." https://www.weather.gov/arx/why_dewpoint_vs_humidity.

———. "Dust Storms and Haboobs." https://www.weather.gov/safety/wind-dust-storm.

———. "How Fog Forms." https://www.weather.gov/lmk/fog_tutorial.

———. "Microbursts." https://www.weather.gov/bmx/outreach_microbursts.

———. "Remembering Delta Flight 191." https://www.weather.gov/fwd/delta191.

Ocean Navigator. "White Squall." June 20, 2013. https://oceannavigator.com/july-august-2013-issue-211-white-squall/.

Pearce, Fred. "How Airplane Contrails Are Helping Make the Planet Warmer." *YaleEnvironment360*, July 18, 2019. https://e360.yale.edu/features/how-airplane-contrails-are-helping-make-the-planet-warmer.

Pride of Baltimore. "History of Pride." https://pride2.org/pride-of-baltimore-ii/history-of-pride/.

Radford, Benjamin. "Bermuda Triangle: Where Facts Disappear." *Live Science*, September 25, 2012. https://www.livescience.com/23435-bermuda-triangle.html.

Saperstein, Saundra, Tom Vesey, Lyle V. Harris, and Mary Jordan. "Four Missing, Eight Saved After *Pride of Baltimore* Sinks." *Washington Post*, May 20, 1986. https://www.washingtonpost.com/archive/politics/1986/05/20/4-missing-8-saved-after-pride-of-baltimore-sinks/99423d30-300c-4104-b43f-cbfde9798e86/.

Shinn, Eugene. "A Geologist's Adventures with Bimini Beachrock and Atlantis True Believers." *Skeptical Inquirer* 28, no. 1 (2004). https://skepticalinquirer.org/2004/01/a-geologists-adventures-with-bimini-beachrock-and-atlantis-true-believers/.

Taylor, Alan. "Photos: Sandstorms Sweep Across Parts of the Middle East." *Atlantic*, May 23, 2022. https://www.theatlantic.com/photo/2022/05/photos-sandstorms-sweep-across-parts-middle-east/629953/.

SOURCES AND ADDITIONAL INFORMATION

Weather Channel. "How Do Lenticular Clouds Form?" Video. https://www.youtube.com/watch?v=u7CVYivLXcY.

Woods Hole Oceanographic Institution. "Currents, Gyres, and Eddies." https://www.whoi.edu/know-your-ocean/ocean-topics/how-the-ocean-works/ocean-circulation/currents-gyres-eddies/.

World Meteorological Organization. "Dust Storm or Sandstorm." https://cloudatlas.wmo.int/en/dust-storm-or-sandstorm.html.

Young, Richard Edward, and Clyde F. Roper. "Bioluminescent Countershading in Midwater Animals: Evidence from Living Squid." *Science* 191, no. 4231 (1976): 1046–48. https://doi.org/10.1126/science.1251214.

7. LIGHTNING

Byrd, Deborah. "What Are Lightning Sprites? How to Photograph Them." EarthSky, July 22, 2023. https://earthsky.org/earth/definition-what-are-lightning-sprites/.

Centers for Disease Control and Prevention (CDC). "U.S. Lightning Strike Deaths." https://www.cdc.gov/disasters/lightning/victimdata/infographic.html.

National Weather Service. "Bolts from the Blue." https://www.weather.gov/pub/lightningBoltBlue.

———. "How Dangerous Is Lightning?" https://www.weather.gov/safety/lightning-odds.

———. "Interesting Facts, Myths, Trivia, and General Information About Lightning." https://www.weather.gov/mlb/lightning_facts.

NOAA National Severe Storms Laboratory. "NSSL Research: Lightning." https://www.nssl.noaa.gov/research/lightning/.

———. "Severe Weather 101: Lightning Basics." https://www.nssl.noaa.gov/education/svrwx101/lightning/.

Occupational Safety and Health Administration (OSHA). "Fact Sheet: Lightning Safety When Working Outdoors." https://www.weather.gov/media/owlie/OSHA_FS-3863_Lightning_Safety_05-2016.pdf.

Plasma Science and Fusion Center, Massachusetts Institute of Technology. "What Is Plasma? https://www.psfc.mit.edu/vision/what_is_plasma.

Scientific American. "What Causes the Strange Glow Known as St. Elmo's Fire? Is This Phenomenon Related to Ball Lightning?" September 22, 1997. https://www.scientificamerican.com/article/quotwhat-causes-the-stran/.

SOURCES AND ADDITIONAL INFORMATION

Wood, Charles. "What Is St. Elmo's Fire?" *Live Science*, November 26, 2019. https://www.livescience.com/st-elmos-fire.html.

World Meteorological Organization. "WMO Certifies Two Megaflash Lightning Records." Press release, February 1, 2022. https://wmo.int/news/media-centre/wmo-certifies-two-megaflash-lightning-records.

8. HURRICANES

Berg, Robbie. "Looking Back at the 2022 Atlantic and East Pacific Hurricane Seasons . . . and What's New from NHC in 2023." National Weather Service, March 8, 2023. https://www.weather.gov/media/nws/IHC2023/Presentations/Berg_NHC.pdf.

Climate.gov. "Historical Hurricane Tracks—GIS Map Viewer." https://www.climate.gov/maps-data/dataset/historical-hurricane-tracks-gis-map-viewer.

Evans, Scotney D., Kenneth Broad, Alberto Cairo, Sharanya J. Pajumdar, Brian D. McNoldy, Barbara Millet, and Leigh Rauk. "An Interdisciplinary Approach to Evaluate Public Comprehension of the 'Cone of Uncertainty' Graphic." *Bulletin of the American Meteorological Society* 103, no.10 (2022): E2214–21. https://doi.org/10.1175/BAMS-D-21-0250.1.

Jacobs, Phie. "Hurricane Otis Smashed Into Mexico and Broke Records. Why Did No One See It Coming? *Science*, October 26, 2023. https://doi.org/10.1126/science.adl5961.

McNoldy, Brian. "2023 'Cone of Uncertainty' Update and Refresher." Tropical Atlantic Update, April 10, 2023. https://bmcnoldy.blogspot.com/2023/04/2023-cone-of-uncertainty-update.html.

National Hurricane Center. "Tropical Cyclone Report: Hurricane Ian." September 23–30, 2022. https://www.nhc.noaa.gov/data/tcr/AL092022_Ian.pdf.

———. "Update on National Hurricane Center Products and Services for 2023." https://www.nhc.noaa.gov/pdf/NHC_New_Products_Updates_2023.pdf.

National Hurricane Center and Central Pacific Hurricane Center. "Definition of the NHC Track Forecast Cone." https://www.nhc.noaa.gov/aboutcone.shtml.

———. "National Hurricane Center Forecast Verification." May 12, 2023. https://www.nhc.noaa.gov/verification/verify5.shtml.

National Ocean Service. "What Is the Difference Between a Hurricane Watch and a Warning?" https://oceanservice.noaa.gov/facts/watch-warning.html.

SOURCES AND ADDITIONAL INFORMATION

NOAA Research. "Thirty Years of Progress in Hurricane Forecasting Since Hurricane Andrew." August 22, 2022. https://research.noaa.gov/article/ArtMID/587/ArticleID/2899/Thirty-years-of-progress-in-hurricane-forecasting-since-Hurricane-Andrew/.

Prager, Ellen. *Dangerous Earth: What We Wish We Knew About Volcanoes, Hurricanes, Climate Change, Earthquakes, and More*. Chicago: University of Chicago Press, 2020.

——. *The Oceans*. New York: McGraw-Hill, 2000.

9. WEATHER FORECASTING AND EXTREME EVENTS

Ballester, Joan, Marcos Quijal-Zamorano, Raúl Fernando Méndez Turrubiates, Ferran Pegenaute, François R. Herrmann, Jean Marie Robine, Xavier Basagaña, Cathryn Tonne, Josep M. Anto, and Hicham Achebak. "Heat-Related Mortality in Europe During the Summer of 2022." *Nature Medicine* 29 (2023): 1857–66. https://www.nature.com/articles/s41591-023-02419-z.

Bishop, Rollin. "How Accurate Is the Farmer's Almanac? Here's What Research Says." *Popular Mechanics*, October 31, 2022. https://www.popularmechanics.com/science/environment/a16962/farmers-almanac-winter-predictions/.

Burke, Patrick C., Alex Lamers, Gregory Carbin, Michael J. Erickson, Mark Klein, Marc Chenard, Jennifer McNatt, and Lance Wood. "The Excessive Rainfall Outlook at the Weather Prediction Center: Operational Definition, Construction, and Real-Time Collaboration." *Bulletin of the American Meteorological Society* 104, no. 3 (2023): E542–62. https://doi.org/10.1175/BAMS-D-21-0281.1.

Copernicus Climate Change Services. "Record Warm November Consolidates 2023 as the Warmest Year." December 7, 2023. https://climate.copernicus.eu/record-warm-november-consolidates-2023-warmest-year.

Farmers' Almanac. "Summer Forecast 2023: Sizzles Return." https://www.farmersalmanac.com/summer-extended-forecast.

First Street Foundation. "The 8th National Risk Assessment: The Precipitation Problem." June 26, 2023. https://report.firststreet.org/8th-National-Risk-Assessment-The-Precipitation-Problem.pdf.

Klesman, Allison. "How Weather Forecasts Are Made." *Discover*, August, 13, 2019. https://www.discovermagazine.com/planet-earth/how-weather-forecasts-are-made.

Livingston, Ian, and Jason Samenow. "A First: Category 5 Storms Have Formed in Every Ocean Basin This Year." *Washington Post*, September 8,

SOURCES AND ADDITIONAL INFORMATION

2023. https://www.washingtonpost.com/weather/2023/09/08/seven-category-5-hurricanes-typhoons-record/.

Maricopa County Department of Public Health. "2023 Weekly Heath Report." November 2023. https://www.maricopa.gov/ArchiveCenter/ViewFile/Item/5734.

Masters, Jeff. "A Record 63 Billion-Dollar Weather Disaster Hit Earth in 2023." Yale Climate Connections, January, 18, 2024. https://yaleclimateconnections.org/2024/01/a-record-63-billion-dollar-weather-disasters-hit-earth-in-2023/.

NASA Earth Observatory. "A Deluge in Greece." September 6, 2032. https://earthobservatory.nasa.gov/images/151807/a-deluge-in-greece.

National Centers for Environmental Information. "U.S. Billion-Dollar Weather and Climate Disasters." https://www.ncei.noaa.gov/access/billions/.

National Oceanic and Atmospheric Administration (NOAA). "Rip Currents." https://www.noaa.gov/jetstream/ocean/rip-currents.

National Severe Storms Laboratory. "Severe Weather 101: Hail Basics." https://www.nssl.noaa.gov/education/svrwx101/hail/.

———. "Warn on Forecast." https://www.nssl.noaa.gov/projects/wof/.

National Tsunami Hazard Mitigation Program. "What Is a Meteotsunami?" https://nws.weather.gov/nthmp/meteotsunamis.html.

National Weather Service. "Heat, Autos, and Safety." https://www.weather.gov/lsx/excessiveheat-automobiles.

———. "National Weather Service Glossary." https://w1.weather.gov/glossary/.

———. "Precipitation Probability." https://www.weather.gov/media/pah/Weather Education/pop.pdf.

———. "Record Setting Hail Event in Vivian, South Dakota on July 23, 2010. https://www.weather.gov/abr/vivianhailstone.

———. "Weather Related Fatality and Injury Statistics." https://www.weather.gov/hazstat/.

———. "What's the Difference Between Sleet, Freezing Rain, and Snow?" https://www.weather.gov/iwx/sleetvsfreezingrain.

NOAA SciJinks. "How Reliable Are Weather Forecasts?" https://scijinks.gov/forecast-reliability.

Prager, Ellen. *Dangerous Earth: What We Wish We Knew About Volcanoes, Hurricanes, Climate Change, Earthquakes, and More.* Chicago: University of Chicago Press, 2020.

Ramirez, Rachel. "Nearly 62,000 People Died from Record-Breaking Heat in Europe Last Summer. It's a Lesson for the US, Too." CNN, July 14, 2023.

SOURCES AND ADDITIONAL INFORMATION

https://www.cnn.com/2023/07/10/world/deadly-europe-heatwave-2022-climate/index.html.

Satterfield, Dan. "Note to CBS: This Is NOT news, It's Make Believe." Dan's Wild Wild Science Journal, August 26, 2013. https://blogs.agu.org/wildwildscience/2013/08/26/note-to-cbs-this-is-not-news-its-make-believe/.

Thomas, Emily A. "Hot Car Fatalities Are a Year-Round Threat to Children and Pets." *Consumer Reports*, September 11, 2023. https://www.consumerreports.org/cars/car-safety/hot-car-fatalities-year-round-threat-to-children-pets-heat-stroke-a2015990109/.

U.S. Environmental Protection Agency (EPA). "Climate Change Indicators: Heat-Related Deaths." https://www.epa.gov/climate-indicators/climate-change-indicators-heat-related-deaths.

Voosen, Paul. "2023 Was the Hottest Year on Record—and Even Hotter Than Expected." *Science*, January 9, 2023. https://www.science.org/content/article/even-warmer-expected-2023-was-hottest-year-record.

World Meteorological Organization. "Storm Daniel Leads to Extreme Rain and Floods in Mediterranean, Heavy Loss of Life in Libya." September 12, 2023. https://wmo.int/media/news/storm-daniel-leads-extreme-rain-and-floods-mediterranean-heavy-loss-of-life-libya.

10. CLIMATE CHANGE

Black, Simon, Ian Parry, and Nate Vernon. "Fossil Fuel Subsidies Surged to Record $7 Trillion." International Monetary Fund Blog, August 24, 2023. https://www.imf.org/en/Blogs/Articles/2023/08/24/fossil-fuel-subsidies-surged-to-record-7-trillion.

Cho, Renée. "You Asked: If CO2 Is Only 0.04% of the Atmosphere, How Does It Drive Global Warming?" Columbia Climate School, State of the Planet, July 30, 2019. https://news.climate.columbia.edu/2019/07/30/co2-drives-global-warming/.

Climate.gov. "Global Climate Dashboard." https://www.climate.gov/climatedashboard.

Crichton, Michael. *State of Fear*. New York: HarperCollins, 2004.

Global Monitoring Laboratory. "Trends in Atmospheric Carbon Dioxide." https://gml.noaa.gov/ccgg/trends/.

Goodell, Jeff. *The Heat Will Kill You First*. New York: Little, Brown, 2023.

SOURCES AND ADDITIONAL INFORMATION

Haigh, Joanna. "The Sun's Influence on Climate Change." Know It Wall. https://www.knowitwall.com/episodes/the-suns-influence-on-climate-change/.

Harris, Alex, and Ashley Miznazi. "Soaring Temps and Record-Breaking Heat Signal Florida's Steamy Future." *All Things Considered* (NPR), July 6, 2023. https://wusfnews.wusf.usf.edu/weather/2023-07-06/soaring-temps-record-breaking-heat-signal-floridas-steamy-future.

Hausfather, Zeke and Piers Forster. " Analysis: How low-sulphur shipping rules are affecting global warming." CarbonBrief.org. March 7, 2023. https://www.carbonbrief.org/analysis-how-low-sulphur-shipping-rules-are-affecting-global-warming/.

Kirtley, David. "The 1970s Ice Age Myth and Time Magazine Covers." ScienceBlogs. https://scienceblogs.com/gregladen/2013/06/04/the-1970s-ice-age-myth-and-time-magazine-covers-by-david-kirtley.

Lindsey, Rebecca. "Climate Change: Global Sea Level." Climate.gov. April 19, 2022. https://www.climate.gov/news-features/understanding-climate/climate-change-global-sea-level.

Mann, Michael E. *The New Climate War.* New York: PublicAffairs, 2021.

Mann, Michael E., and Lee R. Kump. *Dire Predictions: Understanding Climate Change.* New York: DK, 2015.

Mohan, J. E., L. H. Ziska, R. B. Thomas, R. C. Sicher, K. George, J. S. Clark, W. H. Schlesinger. "Biomass and toxicity responses of poison ivy (*Toxicodendron radicans*) to elevated atmospheric CO_2." *Proceedings of the National Academy of Sciences.* 103 no. 24 (2006): 9086. https://doi.org/10.1073/pnas.0602392103

National Academies of Science, Engineering, and Medicine. "Based on Science: Climate Models Reliably Project Future Conditions." October 5, 2021. https://www.nationalacademies.org/based-on-science/climate-models-reliably-project-future-conditions.

National Aeronautics and Space Administration (NASA). "Sea Level." https://climate.nasa.gov/vital-signs/sea-level/.

National Centers for Environmental Information. "U.S. Climate Normals." https://www.ncei.noaa.gov/products/land-based-station/us-climate-normals.

———. "A Warming Earth Is Also a Wetter Earth." https://www.ncei.noaa.gov/news/warming-earth-also-wetter-earth.

National Ocean Service. "2022 Sea Level Rise Technical Report." https://oceanservice.noaa.gov/hazards/sealevelrise/sealevelrise-tech-report-sections.html.

SOURCES AND ADDITIONAL INFORMATION

Peterson, Thomas C., William M. Connolley, and John Fleck. "The Myth of the 1970s Global Cooling Scientific Consensus." *Bulletin of the American Meteorological Society* 89, no. 9 (2008): 1325–38. https://doi.org/10.1175/2008BAMS2370.1.

Piecuch, Christopher G., and Lisa M. Beal. "Robust Weakening of the Gulf Stream During the Past Four Decades Observed in the Florida Straits." *Geophysical Research Letters* 50, no. 18 (2023). https://doi.org/10.1029/2023GL105170.

Prager, Ellen. *Dangerous Earth: What We Wish We Knew About Volcanoes, Hurricanes, Climate Change, Earthquakes, and More.* Chicago: University of Chicago Press, 2020.

Ritchie, Hannah. *Not the End of the World: How We Can Be the First Generation to Build a Sustainable Planet.* New York: Little, Brown Spark, 2023.

Romm, Joseph. *Climate Change: What Everyone Needs to Know.* New York: Oxford University Press, 2016.

Rothman, Lily. "The Real TIME Cover Behind That Fake 'Ice Age' Report." Time.com, May 15, 2017. https://time.com/4778937/fake-time-cover-ice-age/.

Ruiz, Sarah. "2023 was a weird weather year, wasn't it?" Woodwell Climate Research Center, April 23, 2024. https://www.woodwellclimate.org/2023-record-temperatures-weather/.

Santer, Benjamin D., Stephen Po-Chedley, Lilong Zhao, and Karl E. Taylor. "Exceptional stratospheric contribution to human fingerprints on atmospheric temperature." *Proceedings of the National Academy of Sciences* 120, no. 20 (2023). https://doi.org/10.1073/pnas.2300758120.

SeaLevelRise.org. "The Future of Sea Level Rise." https://sealevelrise.org/forecast/.

Skeptical Science. "How Reliable Are Climate Models?" https://skepticalscience.com/climate-models.htm.

Southeast Florida Regional Climate Change Compact. "Unified Sea Level Rise Projections." https://southeastfloridaclimatecompact.org/unified-sea-level-rise-projections/.

Trenberth, Kevin. "How Rising Water Vapour in the Atmosphere Is Amplifying Warming and Making Extreme Weather Worse." *The Conversation*, September 13, 2023. https://theconversation.com/how-rising-water-vapour-in-the-atmosphere-is-amplifying-warming-and-making-extreme-weather-worse-213347.

U.S. Senate Committee on the Budget. "Sen. Whitehouse on Fossil Fuel Subsidies: "We Are Subsidizing the Danger." May 3, 2023. https://www

SOURCES AND ADDITIONAL INFORMATION

.budget.senate.gov/chairman/newsroom/press/sen-whitehouse-on-fossil
-fuel-subsidies-we-are-subsidizing-the-danger-#.
Yin, Jianjun. "Rapid Decadal Acceleration of Sea Level Rise along the U.S. East and Gulf Coasts During 2010–22 and Its Impact on Hurricane-Induced Storm Surge." *Journal of Climate* 36, no. 13 (June 2023): 4511–29. https://doi.org/10.1175/JCLI-D-22-0670.1.

11. THE SUN

Atmospheric Optics. "Inferior Mirage Green Flash." September 26, 2023. https://atoptics.co.uk/atoptics/gf2.htm.
Briggs, Andy. "What Was the Carrington Event, and Why Does It Matter?" EarthSky, October 22, 2023. https://earthsky.org/human-world/carrington-event-1859-solar-storm-effects-today/.
Choi, Charles Q. "What If the Carrington Event, the Largest Solar Storm Ever Recorded, Happened Today?" *Live Science*, March 26, 2024. https://www.livescience.com/carrington-event.
European Space Agency. "Solar Orbiter: The Sun Up Close." https://www.esa.int/Science_Exploration/Space_Science/Solar_Orbiter.
———. "The Sun as You've Never Seen It Before." May 18, 2022. https://www.esa.int/Science_Exploration/Space_Science/Solar_Orbiter/The_Sun_as_you_ve_never_seen_it_before#hedgehog.
Guhathakurta, Madhulika. "How the Coming Solar Maximum Will Impact Us." Interview by Neil deGrasse Tyson, *StarTalk*, August 29, 2023. https://www.youtube.com/watch?app=desktop&v=MH3l9E2y_aI.
Hatfield, Miles. "After Seventeen Years, a Spacecraft Makes Its First Visit Home." NASA News, August 10, 2023. https://www.nasa.gov/feature/goddard/2023/sun/after-seventeen-years-a-spacecraft-makes-its-first-visit-home.
Interrante, Abbey. "NASA Enters the Solar Atmosphere for the First Time, Bringing New Discoveries." NASA News, December 14, 2021. https://www.nasa.gov/feature/goddard/2021/nasa-enters-the-solar-atmosphere-for-the-first-time-bringing-new-discoveries.
My NASA Data. "Solar Eclipses." https://mynasadata.larc.nasa.gov/basic-page/solar-eclipses-background-information.
NASA News. "Future Eclipses." https://solarsystem.nasa.gov/eclipses/future-eclipses/.
———. "Is the Sun Causing Global Warming?" https://climate.nasa.gov/faq/14/is-the-sun-causing-global-warming/.

SOURCES AND ADDITIONAL INFORMATION

———. "NASA HEAT." https://solarsystem.nasa.gov/heat/home/.
———. "NASA Helio Club." September 6, 2023. https://solarsystem.nasa.gov/resources/2992/nasa-helio-club/?category=heat.
———. "Parker Solar Probe." https://science.nasa.gov/mission/parker-solar-probe/.
NASA Parker Solar Probe. "Parker Solar Probe Flies Into the Fast Solar Wind and Finds Its Source." June 8, 2023. http://parkersolarprobe.jhuapl.edu/News-Center/Show-Article.php?articleID=185.
NASA Space Place. "What Causes the Seasons?" https://spaceplace.nasa.gov/seasons/en/.
NASA Space Pl——ace. "What Is Space Weather?" https://spaceplace.nasa.gov/spaceweather/en/.
National Weather Service. "What Causes Halos, Sundogs and Sun Pillars?" https://www.weather.gov/arx/why_halos_sundogs_pillars.
NOAA Space Weather Prediction Center. "Aurora." https://www.swpc.noaa.gov/phenomena/aurora.
———. "Aurora Dashboard (Experimental)." https://www.spaceweather.gov/communities/aurora-dashboard-experimental.
———. "Moderate Flare Event." https://www.swpc.noaa.gov.
———. "Space Weather Phenomena." https://www.swpc.noaa.gov/phenomena.
———. "Sunspots/Solar Cycle." https://www.swpc.noaa.gov/phenomena/sunspotssolar-cycle.
Petersen, Carolyn Collins. "Now We Know How a Solar Storm Took Out a Fleet of Starlinks." *Universe Today*, April 3, 2023. https://phys.org/news/2023-04-solar-storm-fleet-starlinks.html.
Skeptical Science. "Sun and Climate: Moving in Opposite Directions." https://skepticalscience.com/solar-activity-sunspots-global-warming.htm.
University Corporation for Atmospheric Research, Center for Science Education. "Energy from the Sun." https://scied.ucar.edu/learning-zone/earth-system/energy-from-sun.
University of New South Wales, School of Physics. "Mirages and the Green Flash." Physclips. https://www.animations.physics.unsw.edu.au/jw/light/mirages-green-flash-sky-colours.htm.

12. INFORMATION MIXOLOGY

Bey, Justin. "No, Hurricane Irma Is Not a Category 6 storm—There Is No Category 6." CBS News, September 6, 2017. https://www.cbsnews.com/news/category-6-hurricane-myths-of-hurricane-irma/.

Gavin. "Michael Crichton's State of Confusion." RealClimate, December 13, 2004. https://www.realclimate.org/index.php/archives/2004/12/michael-crichtons-state-of-confusion/.

National Hurricane Center and Central Pacific Hurricane Center. "IRMA Graphics Archive: 5-Day Forecast Track, Initial Wind Field and Watch /Warning Graphic." https://www.nhc.noaa.gov/archive/2017/IRMA_graphics.php?product=5day_cone_with_line_and_wind.

National Weather Service. "Detailed Meteorological Summary of Hurricane Irma." https://www.weather.gov/tae/Irma_technical_summary.

Rice, Ken. "No, Cherry-Picked Analysis Doesn't Demonstrate That We're Not in a Climate Crisis." Skeptical Science, October 7, 2022. https://skepticalscience.com/climate_crisis_paper.html.

Scheufele, D. A., and N. M. Krause. Science Audiences, Misinformation, and Fake News. *Proceedings of the National Academy of Sciences* 116, no. 16 (2019): 7662–69. https://doi.org/10.1073/pnas.1805871115.

13. SHOW US THE DATA

American Press Institute. "'Who Shared It?': How Americans Decide What News to Trust on Social Media." March 20, 2017. https://americanpressinstitute.org/publications/reports/survey-research/trust-social-media/.

Besley, John C., and Derek Hill. "Science and Technology: Public Attitudes, Knowledge, and Interest." National Science Foundation Science & Engineering Indicators, May 15, 2020. https://ncses.nsf.gov/pubs/nsb20207/interest-information-sources-and-involvement.

Fioroni, Sarah. "From Institutions to Individuals: How Americans Are Now Looking to Public Figures for News and Information." Knight Foundation, April 25, 2023. https://knightfoundation.org/articles/from-institutions-to-individuals-how-americans-are-now-looking-to-public-figures-for-news-and-information/.

Funk, Cary, Jeffrey Gottfried, and Amy Mitchell. "Science News and Information Today." Pew Research Center, September 20, 2017. https://www.journalism.org/wp-content/uploads/sites/8/2017/09/PJ_2017.09.20_Science-and-News_FINAL.pdf.

Gramlich, John. "What Makes a News Story Trustworthy? Americans Point to the Outlet That Publishes It, Sources Cited." Pew Research Center, June 9, 2021. https://www.pewresearch.org/short-reads/2021/06/09/what

SOURCES AND ADDITIONAL INFORMATION

-makes-a-news-story-trustworthy-americans-point-to-the-outlet-that-publishes-it-sources-cited/.

Liedke, Jacob, and Jeffrey Gottfried. "U.S. Adults Under 30 Now Trust Information from Social Media Almost as Much as from National News Outlets." Pew Research Center, October 27, 2020. https://www.pewresearch.org/short-reads/2022/10/27/u-s-adults-under-30-now-trust-information-from-social-media-almost-as-much-as-from-national-news-outlets/.

Pew Research Center. "Majority of Americans Say They Can Rely on Experts for Science Information." November 9, 2022. https://www.pewresearch.org/short-reads/2022/11/10/americans-report-more-engagement-with-science-news-than-in-2017/ft_2022-11-10_sciencenews_05/.

———. "News Platform Fact Sheet." Pew Research Center, September 20, 2022. https://www.pewresearch.org/journalism/fact-sheet/news-platform-fact-sheet/.

Schwartz, Dan. "The One Group of People Americans Actually Trust on Climate Science." *Atlantic*, February 22, 2022. https://www.theatlantic.com/science/archive/2022/02/weatherman-climate-change/621630/.

Southwell, Brian G., and Karen White. "Science and Technology: Public Perceptions, Awareness, and Information Sources." National Science Board, Science & Engineering Indicators, May 4, 2022. https://ncses.nsf.gov/pubs/nsb20227.

University of California, Berkeley. "Understanding Science 101." https://undsci.berkeley.edu/understanding-science-101/a-scientific-approach-to-life-a-science-toolkit/getting-to-the-source-where-can-i-get-more-information/.

Wilkins, Emily J., Holly M. Miller, Elizabeth Tilak, and Rudy M. Schuster. "Communicating Information on Nature-Related Topics: Preferred Information Channels and Trust in Sources." *PLoS ONE* 13, no. 12 (2018): e0209013. https://doi.org/10.1371/journal.pone.0209013.

INDEX

Advhena magnifica, 11
algae: bloom, 39–40; harmful, 39–40; importance of, 39; and oxygen, 40
Alguita, 12
Alvin, 6–8
American Association for the Advancement of Science (AAAS), 187
American Geophysical Union (AGU), 117, 187
American Meteorological Society (AMS), 187–188
Animal Planet, 2, 4
Atlantis, 67–69
atmosphere, 62, 73, 75, 82–83; carbon dioxide in, 58, 124, 138–140, 143–144, 179; highly charged, 87; and ocean, 189–190; oxygen in, 139, 178; and total solar eclipses, 167; weight at sea level, 7
atmospheric disturbance, 93
atmospheric pressure, 94, 97
atmospheric speed bumps, 75
auroras, 170–172

ball lightning, 80
barracuda, 23
Bermuda Triangle, 62–65, 182; and microbursts, 65; mysterious vanishings in, 63–65
Bettes, Mike, 142
Bimini Road, 67–69
bioluminescence, 69–72
bioluminescent jellyfish, 70
black corals, 54
black ice, 121
blizzard, 122

INDEX

blue jets, 88
Brandon, Peg, 65
Bulletin of the American Meteorological Society, 148
Burgess, George H., 21–22

carbon-12, 144
carbon-13, 144
carbon-14, 144
carbon dioxide: in atmosphere, 58, 124, 138–140, 143–144, 179; causing warming, 138–139; concentrations, 139–141; and plants, 148–149
Carrington, Richard, 171
Carrington event, 171–172
Center for Climate Communication, George Mason University, 187
cherry-picking, 181
climate: changing. *See* climate change; weather and, 136–137
Climate Central, 187
climate change, 61, 78, 135–159; adapting to, 141–142; carbon dioxide and plants, 148–149; carbon dioxide causing warming, 138–139; carbon dioxide concentrations, 139–141; carbon dioxide in atmosphere, 58, 124, 138–140, 143–144, 179; computer-based/numerical climate models, 146–147; "cooling myth," 148; and coral reefs, 57–59; extreme events, 153–156; finding reliable information, 158–159; and global warming, 137–138, 147; and hurricanes, 107;

individuals, role of, 157–158; and intensification of storms, 106; and jellyfish, 36–37; is it a natural cycle, 143–145; ocean currents, 151–153; sea-level rise, 149–151; and sea turtles, 45; and ship emissions, 145; snowpocalypse, 147; solar radiation, 145–146, 175; and Sun, 175; temperatures without thermometers, 142–143; not too late to stop, 156–157; and volcanoes, 145; weather and climate, 136–137
climate-driven drought, 153. *See also* droughts
climate science, 148
computer-based/numerical climate models, 146–147
computer-generated imagery (CGI), 146
condensation trails (contrails), 77–79
"cone of uncertainty," 99–101
confirmation bias, 181
"cooling myth," 148
coral(s): *vs.* coral reef growth, 54–55; growth/reproduction, 52; overview, 50–51; polyps, 50; restoration, 59–61; white, 52–54
coral reefs: and climate change, 57–59; *vs.* coral growth, 54–55; loss of, 49–50; role in protecting shoreline, 55–57; and weather, 55
Coriolis force, 94–95
coronal mass ejections (CMEs), 170–171, 172, 174

INDEX

Crichton, Michael, 148, 181–182
cyclones, and coral reefs, 58

Day After Tomorrow, The, 151
deep-sea exploration, 6–8;
  undiscovered species, 11–12
deep-sea submersibles, 6–7
dinoflagellates (single-celled algae), 71
Discovery Channel, 2
docufiction, 2
dolphins: echolocation in, 40–41;
  sleep, 41; strandings, 42–43
downdrafts, 65, 67. *See also* microbursts
driving through water in road, 128–129
droughts, 79–80, 123, 141, 150, 154, 179
Duobrachium sparksae, 12
dust storms. *See* sandstorms

eagle rays, 45–46
Earth Science Information Partnership (ESIP), 186
echolocation: in dolphins, 40–41;
  whale's, 42
eclipses, solar, 166–169
electrical appliances, and thunderstorm, 85
electromagnetic radiation, 162, 164
El Niño-La Niña, 118, 153
ELVES (Emissions of Light and Very low frequency perturbations due to Electromagnetic pulse Sources), 88

emergency position-indicating radio beacon (EPIRB), 64
evacuations: and hurricanes, 107–109;
  mass, 124; related costs, 96;
  voluntary/mandatory, 108
extreme events, 110–134, 153–156;
  climate change, 153–156;
  prediction of, 125–126
extreme weather, 123–124, 155

Farmers' Almanac, 117–118
Federal Emergency Management Agency (FEMA), 127
fire coral, 28
fire sponge, 28
fireworms, 28
flying fish, 46
fog, 75–76; land, 76; predicting rain, 119; sea, 76; steam, 76
forecast/forecasting: hurricanes, 95–97; intensity, 96, 97; official, 115; rain, 111–112; snow, 121–122
Fransen, Tanja, 119
freezing rain, 120, 122
Fujita, Tetsuya "Ted," 66

geomagnetic storms, 170, 173
Geostationary Lightning Mappers (GLMs), 84, 88
Geostationary Operational Environmental Satellites, 84
Ginsburg, Robert, 176
global climate models, 146
Global Positioning System (GPS), 45, 63–64, 171; and underwater tracking, 10

INDEX

global warming, 147; carbon dioxide causing, 138–139; and climate change, 137–138, 147; defined, 137–138
GOES-R geostationary series, 95
Graham, Ken, 92–93, 102
Granite Mountain Hotshots, 67
Great Barrier Reef, 57
Great Pacific Garbage Patch, 12–14
Great Whirl, 73
green flash: myths surrounding, 165; real, 165–166
greenhouse gases, 138, 148
Guhathakurta, Madhulika (Lika), 162, 172
Gulf of Mexico, 36, 93, 98, 104, 179
Gulf Stream, 64, 72, 98, 152

haboobs. *See* sandstorms
hail, 120–121, 125
hammerhead shark, 20–21
Hanlon, Roger, 43–44
heat, extreme, 57, 155
heat index, 156
heat lightning, 83
heat-related caution, 127–128
Helfman, Gene, 21–22
heliophysics, 160, 162, 172, 175
Hickey, Patrick, 7
Hilderbrand, Doug, 93
Hueter, Bob, 18–19
human pee and jellyfish sting, 31–34
humpback whales, 24–25
Humphries, Susan, 6–7
Hurricane Andrew, 98, 107

hurricane-forecast models, 101–103
Hurricane Harvey, 98, 104, 125
Hurricane Hugo, 98, 107
Hurricane Ian, 98, 99, 105, 106, 125
Hurricane Irma, 179
Hurricane Katrina, 98, 106
Hurricane Lee, 184
Hurricane Opal, 98
Hurricane Otis, 97, 98–99, 123
hurricanes, 90–109; Category 1/Category 2 hurricanes, 103–105; categories, 105, 179; and climate change, 107; cone of uncertainty, 99–101; evacuation, need for, 107–109; forecasts, 95–97; hurricane-forecast models, 101–103; and Indian mound theory, 90–91; needed to create, 93–94; rapid intensification, 97–99; storm surge, 105–106; warning, 99–100; watches, 99–100
Hurricane Sandy, 104, 151
hurricane tracks, 91

Indian burial mounds, 90–91, 181
Indian mound theory, 90–91
Indo-Pacific reefs, 56
Industrial Revolution, 144
information mixology, 176–182
"insta-ice," 120
intensity forecasts, 96, 97
Intergovernmental Panel on Climate Change (IPCC), 158, 187
Irwin, Steve, 23

INDEX

jellyfish, 27, 31–37; bioluminescent, 70; don't attack people, 34–35; as food, 37; nonnative species of, 36; populations, 35–36; reproductive cycle, 35; sting, and pee, 31–34; things to do when stung by, 34
JPSS polar-orbiting satellites, 95
Jurassic Park, 181

Keeling, Charles David, 139
Kelvin-Helmholz waves, 75
Kirk, Michael, 161–162, 171
Kirkpatrick, Barb, 178
Kusche, Larry, 63, 182

land fog, 76
Lawrence Livermore National Laboratory, 144
lenticular clouds, 73–75
lightning, 81–89; causes of, 81–82; chances of being struck by, 89; heat, 83; and rubber-soled shoes, 86; strikes, 83–84; and thunder, 83; without storm, 86–87
Lohman, Ken, 44
Lovell, Jim, 71
luciferase, 70
luciferin, 70

magnetic fields: enormous/chaotic, 169; and sea turtles, 45; and solar minimums, 172; and solar wind, 170
manta, 45–46

Mauna Loa Observatory, 139–140
McNoldy, Brian, 155
megalodons: existence of, 1–4; overview, 2–3
Megalodon: The Monster Shark, 2
Megalodon: The New Evidence, 2
mermaids, 4–6
Mermaids: The Body Found, 2
Mermaids: The New Evidence, 2
meteotsunamis, 133–134
microbursts, 65–66. *See also* downdrafts
mockumentary, 2
Moon, 164, 167–168
Moore, Charles, 12
most dangerous weather, 126–127
Mote Marine Laboratory in Sarasota, Florida, 178
Mummy, The, 79

National Academy of Sciences (NAS), 187
National Aeronautics and Space Administration (NASA), 158, 160, 169, 171, 186, 187
National Hurricane Center (NHC), 92, 96, 100–101, 103, 179
National Oceanic and Atmospheric Administration (NOAA), 4, 9, 10, 92, 113, 158, 186, 187; Climate Prediction Center (CPC), 118; GOES-R series of satellites, 88; Mauna Loa Observatory, 139–140; National Centers for Environmental Information (NCEI), 136; National Hurricane

INDEX

National Oceanic and Atmospheric Administration (NOAA) (*continued*) Center, 99; National Weather Service (NWS), 66, 92–93, 101, 103, 113, 114–115; newest satellites, 98; Ocean Exploration Program, 10; Warn on Forecast technology, 130
National Science Foundation, 186
National Weather Association, 100
National Weather Service (NWS), 66, 92–93, 101, 103, 113, 114–115; Weather Forecast Offices (WFOs), 114
natural cycle, 143–145
Nomura jellyfish, 36
nongovernmental meteorologists, 114
Northern Hemisphere, 94–95, 154, 163–164

ocean: and atmosphere, 189–190; bioluminescence in, 71; glow at night, 69–72; temperatures, and coral reefs, 69; whirlpools in, 72–73
ocean currents: climate change, 151–153; and sea snakes, 28
ocean-observing platforms, 98
ocean temperatures, 3–5, 28, 57–58, 97–98, 107, 123, 140
octopus, 43–44; color-blind, 44; skin, 43
official forecast, 115
Okeanos Explorer, 11–12
Old Farmer's Almanac, 117–118
orcas, 25–27

osmosis, 44
oxygen: and algae, 40; in atmosphere, 139, 178; production by phytoplankton, 178

Packard, Michael, 24
Parker Solar Probe, 173–174
phosphorescence, 69–70
plants: carbon dioxide and, 148–149; pollen-producing, 149; and warming, 141
Popular Mechanics (Rollins), 117
Portuguese man-of-war, 27
Precht, Bill, 56
Pride of Baltimore, 65–66
Pyrodinium bahamense, 71

radiation: electromagnetic, 162, 164; solar, 145–146, 175; ultraviolet, 163
rain: fog predicting, 119; forecast, 111–112; freezing, 120, 122
rapid intensification, 97–99
red corals, 54
reliable information, finding, 158–159
rip currents, 131–132
Rollins, Bishop, 117
rubber-soled shoes and lightning, 86

Saffir, Herb, 103
Saffir-Simpson Hurricane Wind Scale, 103, 179
sandstorms, 79–80
Santer, Ben, 144
Sargasso Sea, 38–39

INDEX

sargassum, 38–39
Sargassum fluitans, 39
Sargassum natans, 39
Satterfield, Dan, 117
scheduling the weather, 116–117
seafloor: charts/maps/images of, 8–9; depictions of, 8–9; mapping deep, 9–10; strange lines on, 10–11
sea fog, 76
sea-level: and coral reefs, 58; rising, 149–151
sea lion *vs.* seal, 47
sea snakes, 28–30
sea stars, 46–47
sea turtles: and climate change, 45; and magnetic fields, 45; navigating ocean, 44–45; and underwater stay, 44
Seidel, Mike, 81
sharks, 16–18; bites, 18; color pattern, 19–20; detection of blood, 18–19; detection of hurricane, 19; important things about, 21–23; killing, 22–23; and shallow water, 20
Sharks: An Animal Answer Guide (Helfman and Burgess), 21
Shinn, Eugene, 67–68
ship emissions, reductions in, 145
Simpson, Bob, 103
siphonophores, 27
sleep: dolphins, 41; whales, 41
Smithsonian Institution, 187
snow forecast, 121–122
Snowmaggeddon, 147
snowpocalypse, 147

Solar Dynamics Observatory (SDO), 173
solar eclipses, 166–169
solar filters, 160
solar flares, 170, 172
solar max, 172–173
solar minimum, 172–173
solar radiation, 145–146, 175
solar storms, 162, 170, 174
Southern Hemisphere, 94–95, 164
space weather, 169–173
sperm whale, 25
Spinrad, Rick, 92
Splash, 4
sprites, 88
starfish. *See* sea stars
State of Fear (Crichton), 148, 181
steam fog, 76
Steingass, Shea, 4–6
St. Elmo's fire, 87
STEREO-A, 173–174
STEREO-B, 173–174
STEREO (Solar TErrestrial RElations Observatory) mission, 173
stingrays, 23–24
stony corals, 51
stony coral tissue loss disease (SCTLD), 58
Storm Daniel, 123
storm surge, 105–106
strandings: dolphins, 42–43; reasons for, 42; whales, 42–43
strange rings/rainbows in sky, 164–165
sulfur emissions, 145
summer heat and the Sun, 163–164

■ 223 ■

INDEX

Sun, 160–175; "burning" ball of gas, 162–163; green flash, 165–166; not the driving force behind climate change, 175; observing, as never before, 173–174; rings and rainbows in sky, 164–165; solar eclipses, 166–169; space weather, 169–173; and summer heat, 163–164; traveling across the sky, 163
sunspots, 117, 170–173
super corals, 60

temperatures: ocean, 3–5, 28, 57–58, 97–98, 107, 123, 140; shown on TV map, 119; without thermometers, 142–143
Terminal Doppler Weather Radar (TDWR), 66
Texas A&M, 178
thermometers, 142–143
thunder and lightning, 83
thunderstorms, 93–94; and electrical appliances, 85; and microburst, 66; supercell, 121; and trees are not good shelter, 84–85; updrafts in, 120
Time magazine, 148
tornado, 129–131; warning, 130–131
total solar eclipses, 167–168
trees are not shelter, and thunderstorm, 84–85
tropical cyclones, 90
tsunamis, 132–133
Twister, 129

ultraviolet radiation, 163
uncertainty/confusion, and hurricanes, 99–101
undiscovered species: and deep-sea exploration, 11–12
unidentified flying objects (UFOs), 93
University Corporation for Atmospheric Research (UCAR), 187
University of Illinois, 117
urine: and jellyfish sting, 31–34
U.S. Air Force Hurricane Hunters, 98
U.S. Forest Service, 187
U.S. Geological Survey (USGS), 186–187
USS *Connecticut*, 9
USS *San Francisco*, 9
USS *Shangri-La*, 71

Vine, Allyn, 7
volcanoes, and climate change, 145

Warn on Forecast technology, 130
weather: and climate, 136–137; extreme, 123–124; most dangerous, 126–127; scheduling, 116–117
weather apps, 112–114
Weather Channel, 81, 96, 110, 142, 192
weather forecasting, 110–134; black ice, 121; extreme impacts/events, 125–126; extreme weather, 123–124; *Farmers' Almanac*, 117–118; fog predicting rain, 119; hail, 120–121; heat-related caution, 127–128; meteotsunamis, 133–134; most

INDEX

dangerous weather, 126–127; NWS, 114–115; official forecast/watch/warning, 115; *Old Farmer's Almanac*, 117–118; partly sunny/partly cloudy, 112; rain forecast, 111–112; rip currents, 131–132; scheduling the weather, 116–117; snow forecast, 121–122; temperature shown on TV map, 119; tornado, 129–131; tsunamis, 132–133; weather apps, 112–114
weather models, 113–114
Weather Prediction Center (WPC), 100, 125
weather radar, 120
weather warnings, 115
weather watches, 115
whales: echolocation, 42; humpback, 24–25; orcas, 25–27; can a person be swallowed by, 24–25; sleep, 41; strandings, 42–43
whirlpools, 72–73
White, Jim, 142, 158
white corals, 52–54
White Squall, 65
Widder, Edie, 69
wildfires, 79, 89, 123–124, 135, 138, 140, 153–154, 179
Woods Hole Oceanographic Institution, 166
World Meteorological Organization (WMO), 93

Yale School of the Environment, 187
Yancey, Paul, 7
Yarnell Hill Fire, 67

zooxanthellae, 51, 53